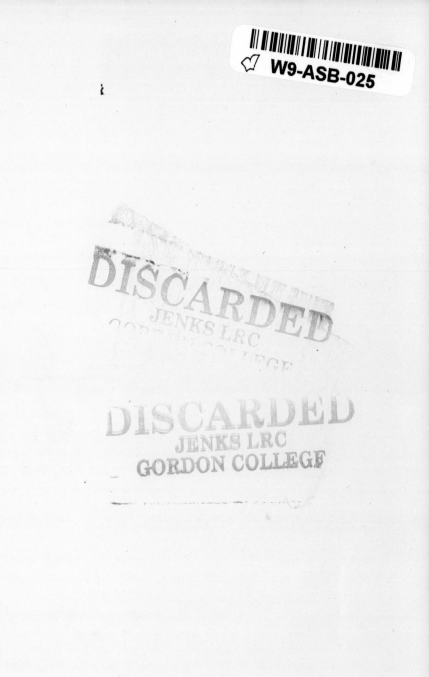

UNIVERSE
DOWN TO
EARTH

Also by
Neil de Grasse Tyson
from
Columbia University Press

MERLIN'S TOUR OF THE UNIVERSE

UNIVERSE
DOWN TO
EARTH

Neil de Grasse Tyson

Columbia University Press

NEW YORK

Columbia University Press
New York   Chichester, West Sussex
Copyright © 1994 Neil de Grasse Tyson
All rights reserved

Library of Congress Cataloging-in-Publication Data

Tyson, Neil de Grasse.
    Universe down to Earth / Neil de Grasse Tyson.
        p.   cm.
    Includes bibliographical references and index.
    ISBN 0–231–07560–X
    1. Cosmology—Popular works.   I. Title.
QB982.T97 1994
523.1—dc20   93–32259
                                                              CIP

∞

Casebound editions of Columbia University Press books are printed
on permanent and durable acid-free paper.

Printed in the United States of America

c 10 9 8 7 6 5 4 3 2 1

For
Alice Mae Young

Contents

Preface

It is not the intent of this book for you to learn the latest scientific discoveries in astronomy; unfortunately, many such books are out of date even before they make it to press. It is not the intent of this book to train you to recite science facts at a cocktail party; there is no shortage of books with that as a mission. Nor is it the intent of this book to amuse you with anecdotal stories on science personalities.

Contained in these pages is the consequence of my attempts to convey science concepts to working adults, college students, and to the average person. There is a greater need than ever for educators and the community of research scientists to find conduits through which the principles of science can be conveyed to the lay audience. It is my experience that success requires not only the proverbial translation from jargon to lay terms—it also requires that an instructor develop a sensitivity to those tangled mental roadways that may confuse a person's attempt to understand a concept. Only then can one claim to communicate with an audience rather than simply lecture to it.

It is not always obvious what it takes to remember a morsel of trivia, a physical principle, or a profound idea. But it is well known that if you seek intellectual enlightenment, then rote memory is not as important as insight. Insight is what remains after you have forgotten all the details. Yes, in this book you will find some cosmic factoids and an anecdote or two. But the book's objective is to convey ideas that etch deeply enough on the mind so that the concepts are not just remembered—they are absorbed into one's intuition.

Often, deep understanding comes from analogy—not poetic analogy but physical or household analogies drawn from one's own path through life. Such analogies are common throughout these pages, in my earnest attempt to bring at least a part of the universe down to Earth.

Acknowledgments

The chapters that appear in *Universe Down to Earth* were derived, in part, from special topic lectures that I delivered to undergraduates at the University of Texas at Austin, the University of Maryland, and Columbia University. Many of the essays have appeared, in slightly abbreviated form, as feature articles in *Stardate Magazine* between 1983 and 1993.

It is a pleasure to recognize the following educators and scientists whose pedagogical visions have helped to shape those of my own: Mark Chartrand III, Fred Hess, Alexander Taffel, Carl Sagan, Isaac Asimov, Frank Bash, Joseph Patterson, and John Archibald Wheeler. I also thank Ann Rae Jonas, the former editor of *Stardate Magazine,* for her continued interest in my work and her efforts to ensure that I always say what I mean.

Several perspectives and occasional assorted facts that appear in *Universe Down to Earth* evolved from specific conversations held with certain colleagues, friends, and relatives: Richard Binzel, Bohdan Paczynski, Robert Lupton, George Wallerstein, David Spergel, Brian Marsden, Alice Mae Young, Lynn Tyson, Lauren Vosburgh, Richard Vosburgh, Jeffrey Rayner, Sangeeta Malhotra, Richard Joyce, and Charlie Caldwell. I am particularly grateful to Robert Lupton for his careful reading and helpful suggestions on an early version of the manuscript.

List of Figures

UNIVERSE
DOWN TO
EARTH

PART ONE

·

Methods

of

Science

· 1 ·

A Sentimental Journey to the Googolplex

The methods of science extend their deepest roots in the origin of numbers. We owe it to the "number line" to give due attention to all numbers big and small. As one who studies the universe, however, it should come as no surprise that I have slanted the coverage toward the big ones.

If somebody asks you, "How big is a billion?"—what would be your response? If you didn't have much to say, then one could infer that you are probably not an astronomer. Of all the natural and physical sciences, astronomers dominate the "big numbers" market. Indeed, it is nearly impossible to conduct a conversation with an astronomer without greeting numbers that contain more zeros than you would bother to count. This chapter may help you appreciate the evolution of this phenomenon.

ᘓᕲ

*I*t is generally agreed among historians that economics played a role in the birth of mathematics. For example, if I breed chickens and you breed sheep, and I want some of your sheep, it would be natural for us to swap chickens for sheep. But first we must answer the question: How many chickens equal one sheep? This seemingly simple question in fact requires the invention of a logical scheme for counting. Some of the earliest evidence for the ability to count comes to us on a 30,000-year-old wolf bone excavated in eastern Europe that was deeply etched with fifty-five notches in groups of five. But the ability to count, undeniably a sophisticated concept, is still not sufficient to deal with all our problems. Suppose I have only five chickens, but you think a sheep is worth ten. I cannot afford to "buy" a whole sheep. With this predicament, a revolutionary concept of numbers is needed to help us consummate our trade: the concept of a half a sheep. But we need not stop with only half a sheep. Suppose my unit of barter is not a chicken but a pea from my vegetable garden. Certainly five peas will buy much less than half a sheep—perhaps only one-thousandth of a sheep. Apart from the logistical difficulty of actually trading a fraction of a (living) sheep, it is clear that advanced barter requires one to be comfortable with numerical quantities borrowed from the world of fractions.

For most of the five thousand years of recorded history, little scientific significance was attached to extremely small quantities. It was not until the late 1600s, when the Dutch naturalist Anton van Leeuwenhoek introduced the microscope to the world of biology. With the desire to measure precisely the sizes of cells, protozoa, and bacteria, there proliferated tiny fractions of the measuring unit such as one-thousandth, one-millionth, and one thousand-millionth.

Meanwhile, astronomers actively explored in the opposite direction with the help of Galileo Galilei's introduction of the telescope to the world of astronomy in the year 1610. The telescope heralded a new scientific era that allowed astronomers to worry about, and subsequently estimate, the size of objects in the universe and their distance from the Earth. Equally bulky numbers then emerged, such as one million billion, one billion trillion, and one trillion trillion.

Astronomers and biologists alike were faced with the same problem: How can one talk cleanly and precisely about extreme quantities of the universe without polluting the conversation with countless "–illions" or "–illionths"? And European travelers need not be reminded that a billion in England and most of Europe is a thousand times bigger than a billion in the United States and France. This numerical dilemma was compounded in the early twentieth century after atoms and subatomic particles were discovered by the physicists J. J. Thompson, Ernest Rutherford, and James Chadwick at the Cavendish Laboratory of the University of Cambridge.

What follows is a quick journey through some awkward "–illionths" and "–illions" that once bedeviled the scientific community.

When measuring in parts of a meter:

one-thousandth	The fraction one over one with three zeros. This is the approximate radius of a peppercorn.
one-millionth	The fraction one over one with six zeros. The head of a human sperm typically has this radius.
one-billionth	The fraction one over one with nine

zeros. This is a common radius for the tiniest bacteria.

one-trillionth The fraction one over one with twelve zeros. Fifty-three of these is the classical radius of the hydrogen atom.

one-quadrillionth The fraction one over one with fifteen zeros. About three of these will get you the classical radius of the electron.

one-quintillionth The fraction one over one with eighteen zeros. This slice of a sheep would not buy much in any economy.

When simply counting, we get:

one thousand Written as a one followed by three zeros. This is about the number of times per minute that Earth is struck by lightning.

one million Written as a one followed by six zeros. When counting people, there are about eight of these piled into New York City.

one billion Written as a one followed by nine zeros. If you never slept, it would take you thirty-two years to count this high. And cows are dismayed to learn that, when last checked, McDonald's hamburger food chain has sold more than one hundred of these. Laid end to end, this many hamburgers would go around the Earth two hundred times, and, with what remains after you have eaten a few, would bridge three round trips to the Moon.

one trillion Written as a one followed by twelve zeros. This is about how many seconds of time have passed since the Neanderthal roamed Europe, Asia, and northern Africa.

one quadrillion   Written as a one followed by fifteen zeros. When the human population reaches this number, then everybody will have to stand upright in order to fit on the Earth's surface.

one quintillion   Written as one followed by eighteen zeros. This is the sum of all sounds and words ever uttered since the dawn of the human species. The tally includes congressional debates and filibusters. Coincidentally, this is also about the number of grains of sand on an average beach.

one sextillion   Written as one followed by twenty-one zeros. This is the estimated number of stars in the universe.

one bezillion   In spite of what your friends may tell you, this number does not really exist.

The scientific community, unhappy with such awkward terminology, needed a more elegant method of numerical organization. Hence, a sensible system of prefixes was formalized by the International Union of Pure and Applied Physics to be used in conjunction with the metric system. With such a scheme, physical quantities are described in units of thousands so that every three zeros appended to a number yields a new prefix. Additionally, scientific notation was introduced to prevent writer's cramp if, for some reason, you chose to write out the number. For example, a large number like 2,000,000 (two million) can be written in scientific notation as 2.0×10^6, where the "6" in the "10^6" tells you how many places the decimal hops to the right. For a tiny number such as .000002 (two-millionths), scientific notation represents it as a 2.0×10^{-6}, where the "-6" in the "10^{-6}" tells you how many places the decimal moves to the left.

The officially accepted prefixes are listed below.

yotta- 10^{24}	giga- 10^{9}	deci- 10^{-1}	pico- 10^{-12}
zetta- 10^{21}	mega- 10^{6}	centi- 10^{-2}	femto- 10^{-15}
exa- 10^{18}	kilo- 10^{3}	mili- 10^{-3}	atto- 10^{-18}
peta- 10^{15}	hecto- 10^{2}	micro- 10^{-6}	zepto- 10^{-21}
tera- 10^{12}	deka- 10^{1}	mano- 10^{-9}	yocto- 10^{-24}

From what appears to be a melodic quartet in pentameter, one simply locates the correct prefix and appends it to whatever quantity is measured. Some common examples are centimeter (one-hundredth of a meter), kilogram (one thousand grams), and megahertz (one million hertz).

A romp through the scientific offerings of the universe can occasionally tempt you to invent new units that will make measurement simpler for the intended task. This, of course, has already been done for the smallest to the largest length scales. The branch of physics known as quantum mechanics dictates that the structure of space itself is discontinuous on scales of what is called the "Planck length," which is about 1.6×10^{-33} centimeters. The role of the German physicist Max Planck, in the dawn of quantum mechanics, is discussed further in chapter 8. Atomic distances and wavelengths of visible light are commonly measured in units of "Ångstroms," which is defined to be 10^{-8} centimeters. Yellow light has a wavelength of about 5,000 Ångstroms.

Distances among planets in the solar system are conveniently measured in "astronomical units," which is defined to be the average distance between Earth and the Sun— about 93 million miles. On this scale, for example, the average distance of Pluto from the Sun is just under 40 astronomical units. The distance that light travels in one year is enormous. This is the famous "light-year," which is about 5.8 trillion miles. It forms a convenient yardstick to measure distances between the stars. The nearest star to the Sun— Proxima Centauri—is about 4.1 light-years away.

For obscure historical reasons, most astronomers use the "parsec" rather than the light-year as the yardstick of choice. One parsec equals 3.26 light-years. There is no widely used unit of distance that is larger than the parsec, although one could, in principle, define the size of the entire universe (about 15 billion light-years in diameter) as a new yardstick, but it would not be very useful—what else could you find to measure with it?

In any adopted set of units, there is no doubt that astronomers monopolize the big numbers. But the biggest number of them all—the one that signifies the physical limit of measurable nature—is a very clean, compact-looking number within which all of astronomy is contained:

$$10^{81}$$

This seemingly unremarkable quantity actually represents the estimated number of atoms in the universe, yet it has no name. How about totillion? If you are worried over the fact that each of these atoms also contains discrete *sub*atomic particles that can in turn be counted, don't worry too much. Over 90 percent of all atoms in the universe are hydrogen atoms. Hydrogen in its most common form contains no neutrons—only a single proton and a single electron. For a better estimate of the total number of atoms and subatomic particles in the entire universe, we should simply double our newly named number: 2×10^{81}

Does this mean that we cannot discuss numbers bigger than 2×10^{81} ? Certainly not. We must simply remember that such numbers have no relationship with physically countable quantities in nature. Let's take 10^{100}, for example.

It is a one followed by one hundred zeros. This rounded, neat-looking number, which is ten quadrillion times larger than the number of atoms in the universe, actually has a name. It was christened a "googol" by a nine-year-old nephew of the mathematician Edward Kasner. Though it is a worthy, even lovable, number, it is not my favorite. That distinction goes to the number ten raised to the googol power:

$$10^{(\text{googol})} = 10^{10^{100}}$$

which has the immortal name of the "googolplex." This number was originally supposed to be a one followed by as many zeros as it would take for someone to get tired of writing them. Since different people obviously get tired at different rates, the googolplex was redefined in terms of the googol. Thus, a googolplex is a number so big that it cannot be written without the aid of scientific notation. It has more zeros than can fit in the universe. Actually, this should not be surprising since a googolplex is a one followed by a googol zeros, and a googol is a number bigger than the sum of all particles in the universe. Even if you could write your zeros small enough to place one on every existing atom, the googolplex still could not be written out in the space of the universe. It is sobering to see the dynamic range of astronomy humbled by the imagination of a nine-year-old just as it is enlightening to realize that one's imagination can extend beyond the limits of astronomical perspectives.

Just for the record, there exists yet another named number that dwarfs even the googolplex:

$$10^{10^{10^{34}}}$$

This is known as Skewes's number, which gives mathematicians information about the distribution of prime numbers. Skewes's number can also be discussed abstractly even though it obviously has no measurable application to nature. For example, the mathematician G. H. Hardy pointed out that if the entire universe were a giant cosmic chessboard, and the interchange of protons between any two atoms were legal, then Skewes's number would represent the total possible number of moves!

Is this the whole story of science and numbers, or can there be another level of investigation? One cloudy night, when I had nothing better to do, I decided to look more closely at our revered international system of metric prefixes.

In these days of inflated modifiers, a thank-you note might get more attention if you signed it "Thanks $\times$ 10^9," provided, of course, you really do mean "Thanks a billion."

What happens if you are 10^{-6} biologist? That must make you a microbiologist.

How about if you decided one day to read a copy of 2×10^3 Mockingbird? That would of course be Harper Lee's undiscovered classic novel, *Two Kilo Mockingbird*.

If you have ever played 10^{-12} boo with a child, then it was probably the metric version called Pico-boo.

If the facial beauty of Helen of Troy was sufficient to launch a thousand ships, then the 10^{-3} Helen, better known as the milli-Helen, must be the amount of beauty required to launch just one ship.

Suppose you owned 10^1 cards? That of course would be your own personal deka-cards.

What happens if you live in a 10^6 lopolis? This is none other than a megalopolis.

And finally, if you just had a 10^{-2} mental journey to the googolplex, what kind of journey was it?

· 2 ·

The Structure of Science

*Science consists in discovering the frame and operations
of Nature, and reducing them, as far as may be, to
general rules or laws—establishing these rules by
observations and experiments, and thence deducing the
causes and effects of things.*

—*Sir Isaac Newton*

⟋⟍

Scientific Method

*The most remarkable discovery made by scientists is
science itself.*

—Gerard Piel

Nearly every century since the dawn of recorded history can
boast a list of scientific achievements, but no list is as extensive or has had as much impact on world order as that of
the twentieth century. For example, quasars, the expanding
universe, the proton, the neutron, the theory of relativity,
quantum mechanics, the airplane, vitamins, vaccines, DNA,
and nuclear bombs were all discovered or invented after the
year 1900. We often take this knowledge of the physical
world for granted, as though it were inscribed on a cosmic
tablet in the sky. But most scientific discoveries come from
sweat, perseverance, and insight. We may attribute this
brow-raising success of twentieth-century science to three
sources: (1) the unprecedented number of scientists in the
world today, (2) the development of laboratory equipment
that makes available to the human senses views of nature
that would otherwise go unnoticed, and (3) the widespread
and successful application of the scientific method.

The scientific method is responsible for the discoveries
and the predictions that shape our understanding of the
universe. It is the guiding principle of nearly all science.
Whether you know it or not, you are exposed to the scientific
method almost daily. In its most basic form, the scientific
method demands that an experiment or observation be conducted to test a hypothesis. We find a colloquial version of
the scientific method with the residents of Missouri, who (it

is rumored) have a statewide compulsion to say "Show Me!" if you approach them with a hard-to-believe claim. On a national scale, by joint mandate of the Federal Communications Commission and the Federal Trade Commission, a television advertisement must illustrate the scientific method to substantiate any claim that is made about an advertised product. This is why stains are lifted, ring-around-the-collar is removed, paper towels become soaked, excess stomach acid is absorbed, and headaches go away—all during the commercial.

Two millennia ago, in the days before television commercials, there lived a Greek philosopher named Aristotle whose writings about the physical and biological world were, on occasion, strikingly devoid of the scientific method. To be fair to Aristotle, the scientific method had not yet been formally developed. It emerged in the more familiar and appreciated form we know today principally through the work of Galileo Galilei and Sir Isaac Newton in the seventeenth and eighteenth centuries. But Aristotle was so influential that he may have delayed the advent of the scientific method by several centuries. Aristotle's method would often be just to think about the world and declare what ought to be true. The truth was presumed to be self-evident as inferred from the expectation of a philosophically perfect universe. It was self-evident that a 100-pound iron ball would fall to the ground one hundred times faster than a one-pound iron ball. It was self-evident that the source of all human feeling and emotion was the heart. It was self-evident that Earth was in the center of the universe. It became quite evident that Aristotle was wrong.

Some of Aristotle's claims are easier to test than others. You are not likely to attract volunteers who will let you remove their hearts to see if they retain their emotions, but anybody can drop two objects to the ground. If you perform this experiment yourself, you will discover that a big rock and a little rock will fall at exactly the same rate. This experiment, of course, does not require rocks. It will also work in the privacy of your home with a bowling ball and a billiard ball, or a bowling ball and a golf ball, or a bowling ball and a marble, or even a bowling ball and a feather (if you removed all the air from your room), but the downstairs neighbors are not likely to appreciate these experiments.

Over the centuries, one by one, many of the Aristotelian claims were abandoned. It is astonishing to realize that until Galileo performed his experiments on the acceleration of gravity in the early seventeenth century, nobody questioned Aristotle's falling balls. Nobody said, "Show Me!"

When we believe that we understand nature on some level, we can then attempt to predict events. Scientific theories are ideas that explain some of what is already known and predict the future behavior of some of what is not known. Some of the first of their kind were the three "laws" of planetary motion published in 1609 and 1619 by the German astronomer Johannes Kepler. His third law contains a mathematical equation that predicts the orbital period of a planet if you know its average distance to the Sun. It was by no means self-evident that such a relation should exist. Kepler labored for ten years over extensive and accurate data on planetary positions that were willed to him by Tycho Brahe, a well-funded and flamboyant observational astronomer from Denmark. The result is a triumph of the scientific method, where the importance of data and the application of mathematics to a physical problem became a tradition that was carried forth to the laboratories of modern science.

Paths of Science

Natural science does not simply describe and explain nature;
it is a part of the interplay between nature and ourselves; it
describes nature as exposed to our method of questioning.
—*Werner Heisenberg*

Now there is a sobering thought. Heisenberg deftly declares that we are not passive, detached observers of natural law. We are active participants who unravel the universe in the context of our intellectual and experimental limitations.

A creative scientific mind is one that combines formal training with personal insight to achieve a discovery. Upon becoming a scientist, you bring with yourself a unique point of view. It is amended and refined throughout one's career, although it is not necessarily true that the best scientists should be the ones with the longest path of life. Sir Isaac Newton, Charles Darwin, Albert Einstein, Werner Heisenberg, and Wolfgang Pauli—towering greats in the history of science—all made their most profound contributions before age thirty. Nor is it necessarily true that early training in a chosen scientific discipline best prepares you for later discovery in that discipline. Martin Harwit, in his book *Cosmic Discovery*, notes that after 1945 about 70 percent of the important astronomical findings were made by scientists not originally trained as astronomers. The best examples of discovery come from those who invent a new piece of hardware such as a detector, or from those who introduce a new telescope design. The utility of such a device is often immediately apparent to all, but it is the inventor who often gets to use it first.

When you read newspaper accounts of scientific discovery, you are not always told that there were countless other research paths that were scientifically sound but produced unfruitful, misleading, or wrong results. In some cases this realization can lead to disillusionment. The Nobel Prize-winning physicist Wolfgang Pauli, in the predawn of quantum mechanics, lamented,

> At the moment physics is terribly confused. In any case, it is too difficult for me, and I wish I had been a movie comedian or something of the sort and had never heard of physics.

One does not normally hear about the daily failures on the path to discovery. In more blunt terms, the public reads about the hits and not the misses. There is a scene in the film *The Wizard of Oz* where Dorothy (and Toto too) come to a fork in the yellow brick road. The Scarecrow, after what appears to be random arm-swingings, points to what is later revealed to be the correct path to Oz. Once again, the audience is treated to a hit and not a miss. I often wonder how boring the film would have been if Dorothy and Toto missed and had taken the wrong path. Not only would Dorothy have been lost in her dream, she and her little dog wouldn't have met such charming characters as the ugly bad witch, the ill-mannered talking trees, and the winged monkeys.

Science is replete with examples of time devoted to wrong paths. But this does not mean that we should abandon science or the scientific method. As early as the seventeenth century, Francis Bacon noted that truth emerges more readily from error than from confusion. Some spectacularly wrong concepts include Ptolemy's earth-centered (geocentric) universe, the caloric theory of heat (where heat is taken to be a physical substance), and the luminiferous ether (which was thought to permeate all of space as an omnipresent medium for light to propagate). From these misguided ideas grew the Copernican sun-centered (heliocentric) model of the solar system, thermodynamics, and Einstein's special theory of relativity, which led later to his general theory of relativity. In a more earthly example we find that cows rather than lions are domestic farm animals because (among other reasons) while a cow is content to eat grass, a lion is content to eat the farmer. Yes, in science and in animal husbandry, wrong paths can be the source of their own undoing.

One's point of view, when merged with one's path of life, provides the framework within which steps are taken in the name of scientific progress. As we have seen, a point of view can sometimes be skewed. What some people may consider to be an insightful point of view may even tell more about a person's bias than about a scientific truth. For example, in the 1940s and 1950s, the now-defunct steady state theory of the universe was developed and held to be aesthetically more fulfilling than the oscillating universe. (At the time, there were not enough data to distinguish one model from another.) In the steady state theory, space is infinite and all regions of the universe look the same, on average, for all of time. The oscillating model, however, holds that our expanding universe will one day recollapse and then explode into existence again. If cosmological theory were dominated by women, who are no strangers to cycles, how can we know for sure that we wouldn't then be told that the oscillating universe is the more aesthetically fulfilling alternative? Another example comes from the field of mathematics, where there is a theorem that colloquially translates, "You cannot comb the hair on a bowling ball." Aside from the complication that most bowling balls have no hair, the theorem means that if a bowling ball did have hair, and if you tried to comb it, there would be at least one spot on the ball where the hair would not know which way to lay. Clearly, none of these mathematicians had Afros, because to "comb" an Afro is to pick it straight away from the scalp. If bowling balls had Afros, then yes, they could be combed *without* violation of mathematical theorems.

Sometimes a naive observation can lead to misnomers and misconceptions. Historically, when very bright new stars appeared in the sky, each was promptly declared to be a nova, from the Latin *novus*, meaning "new." Stars that were even brighter were each called a supernova. The ensuing paths of research unfolded with the misguided assumption that these stars were indeed new. We now know these objects to represent the explosive end stage of a star's life. A more appropriate pair of terms might be "mortem" and "supermortem."

In other examples, there is a deficiency in the vocabulary. The words "particle" and "wave" have a worthy, independent existence. But when the behavior of photons is discussed, conceptual problems emerge because photons are accurately and legitimately described as waves *and* particles at the same time. They are genuine "wavicles." A bias in the minds of many pre-quantum mechanics physicists prevented their acceptance of this unorthodox view. Similar trouble emerged with the words "space" and "time." After 1905, when Einstein first introduced special relativity, the marriage was inevitable; space and time became "space-time." As noted by the Russian mathematician Hermann Minkowski in 1908,

> Nobody has ever noticed a place except at a time, or a time except at a place. . . . Henceforth, space by itself, and time by itself, are doomed to fade away into mere shadows, and only a kind of union of the two will preserve an independent reality.

Those trapped with conceptual blinders were the last to be convinced of the merits of the new physics.

Wrong analogies also form skewed paths. Between 1609 and 1619, Johannes Kepler was in search of his third law of planetary motion. His mathematical training led him to be fully aware of the five regular geometric solids: the tetrahedron (four sides of identical triangles), the cube (six sides of identical squares), the octahedron (eight sides of identical triangles), the dodecahedron (twelve sides of identical pentagons), and the icosahedron (twenty sides of identical triangles). What makes these solids unique among all possible shapes is that for each solid, all edges have exactly the same length. These are the *only* five solids that qualify. No one will doubt that these solid shapes are intriguing. But Kepler was so intrigued that he wasted ten years of his life in pursuit of a mystical-philosophical assumption: that the geometry of these five solids and the five distances between the orbits of the then-known planets (Mercury, Venus, Earth, Mars, Jupiter, and Saturn) were somehow related.

A more modern example brings us to Einstein. He was philosophically uncomfortable with the new field of quantum mechanics. In quantum mechanics, subatomic nature can be described only statistically, while Einstein believed nature to be deterministic on all scales. Einstein is often quoted as saying, "God does not play dice with the universe." What he really means is "It would fit better into my conception of the universe if God did not play dice." Unfortunately for Einstein, with regard to quantum mechanics, God *does* play dice. And unfortunately for Kepler, planets' orbits have *no relation* to geometric solids.

Importance of Data

First get your facts; and then you can distort them
at your leisure.
—Samuel Clemens

Data are bits of information obtained from a physical process about which models and theories are constructed. All data are subject to what are called "measurement errors." This is a most unfortunate term because when we discuss the errors in our data with the general public they often wonder if we constantly make mistakes. A less confusing term would be "experimental result range" (although its acronym, ERR, is not much of an improvement over "error"). When a scientist measures a phenomenon several times, it is *expected* that the results will vary from the true value of what is measured. It is *presumed* that the average value of the measurements indicates this true value. But an average alone is not especially informative. The average of 0 and 10 is 5, but 5 is also the average of $4\frac{1}{2}$ and $5\frac{1}{2}$. The tightness of the measurements around the average is an important quantity called the "deviation," which can lend confidence to the average that is computed from the data. Clearly, the 5 is a more telling average for the second pair of numbers than for the first pair of numbers. A common measure of the tightness of the deviation is the sum of the square of each deviation divided by the number of measurements. The square root of this number is called the "standard deviation," which is one of the simplest and most widely used statistical descriptions for the accuracy of data. If you seek a more complicated test, there are professional journals in the field of statistics that are devoted entirely to methods of data analysis. But there is often healthy skepticism of scientific results that require exotic statistical tests for their believability. As demonstrated through the history of experimental science, results that can be justified through simple statistical tests are more likely to be correct. With somewhat exaggerated skepticism, the esteemed New Zealand physicist Lord Ernest Rutherford noted, "If your experiment needs statistics, you ought to have done a better experiment."

A scientific study is not the truth simply because it is a scientific study. Science is conducted by human beings who are not immune from errors caused by blunder, personal bias, or even fraud. Data may occasionally contain errors of contamination. For example, let's say you invent an apparatus that counts the number of raindrops that fall on your lawn. But your neighbor, who lives fifty feet away, runs a lawn sprinkler that sprays one hundred feet. The data you obtain will say more about your neighbor's lawn habits than the precipitation over your house. A genuine example of contamination occurred in the early 1970s. Telescopes near major cities tended to show specific and identical sodium features in the spectra of all stars. What seemed, at first, to be a cosmic sodium presence was ultimately determined to be scattered light in Earth's atmosphere from the new sodium vapor street lamps that are now common in urban areas.

The way to obtain high-quality data is to design a high-quality, blunder-resistant, bias-resistant experiment. The most secure experiments are the ones where any result would be interesting, even a null result. An example of an interesting null result would be if you went out to the side of your neighborhood's busiest street to count the number of Lamborghini Countachs (or Ford Model Ts) that pass per hour. If you get zero for every hour of every day for an entire year, then you have every right to postulate that there are no Lamborghini Countachs in your neighborhood. You have thus made a valid and somewhat interesting conclusion from a null result.

Arguably the most important null-result in the history of physics was the 1887 experiment by the American physicist Albert Michelson and the American chemist Edward Morley. They sought to determine the change in the speed of light when measured at different angles to Earth's orbital motion around the Sun. To their amazement, as well as to the amazement of the scientific community, the measured speed of light remained constant in all directions. This humble null-result helped to bury the long-held concept of the luminiferous ether and laid one of the foundations for Einstein's principles of relativity.

To collect data is often a long and arduous process. In the end, we don't look for the right or wrong answer, we look for consistency. It is not whether one study finds evidence in support of a claim, it is whether all studies find evidence in support of a claim. We ask, "Is Dr. Aster's research results consistent with those of Dr. Roid's?" Equivalently, we can ask, "Are the data from Dr. Aster's experiment within the range of data from Dr. Roid's experiment?" If the answer is "No," then we ask, "Were their experiments or observations of similar design?" or, "Might there be errors other than errors of measurement?" Scientists will, in general, pursue those topics that are not yet resolved because these are the most exciting research areas. When the data from different researchers are consistent with each other and when a theorist helps to provide a predictive understanding for the measured phenomenon, then the boundary of human understanding of the universe has grown.

Mathematics

There is something fascinating about science. One gets such wholesale returns of conjecture out of such a trifling investment of fact.

—*Mark Twain*

If you happen to be in Katmandu, and your Nepali dialect is rusty, and you wish to buy a snack, and you remain clueless even after the merchant tells you that it costs "nau rupya or pachas pice," then just motion for the price to be written. The merchant will likely use the most widely accepted symbols of them all: 0–1–2–3–4–5–6–7–8–9, and write *9.50*. You will then both be happy. This universality of numbers, and of mathematics in general, provides an international conduit through which you may buy snacks in exotic places. And since mathematics is the language of the physical sciences, it also allows scientists to communicate discoveries throughout the world. We know that skyscraper office buildings, the steel icons of urban centers, could not have been realized without the tandem development of appropriate construction tools and machinery. Such is the union of science and mathematics. While science was being discovered, mathematical branches and their tools were being invented. Some noteworthy examples follow.

The Greek mathematician Euclid laid the foundation for two millennia of plane geometry in his book *The Elements*, which appeared circa 300 B.C. Armed with these tools, another Greek mathematician, Eratosthenes, measured the circumference of Earth. (It should now come as no surprise that the term "geometry" derives from the two Greek words that translate to "earth-measurement.")

You may remember Cartesian coordinates from high school. René Descartes, a celebrated mathematician of the seventeenth century, introduced the concept of coordinate systems, which is the foundation of graphical solutions to problems that have been mathematically formulated.

Nonplane geometry is the fabric of curved space-time in Albert Einstein's relativity theory. It is, with inverted respect, called *non*-Euclidean geometry, although it was developed eighteen hundred years later than Euclid's geometry, principally through the work of Nikolai Lobachevski in the late sixteenth century.

In the eighteenth century, the brilliant German mathematician Karl Gauss invented the powerful statistical method called "least squares" to compute with unprecedented precision the orbit of the asteroid Ceres from incomplete data available to him.

The Scottish nobleman John Napier invented logarithms and the "slide rule" in the late sixteenth century, which simplified computations among scientists for more than three hundred years. It was not until the 1970s that the hand-held calculator rendered slide rules obsolete. Yes, among researchers at the major computation centers of the world, the side-of-the-thigh slide rule holsters were rapidly replaced by zip-closable calculator pouches that were donned at the hip.

With the available mathematics of the mid-1600s, Sir Isaac Newton could not easily demonstrate why planets orbited in ellipses. By 1666, he summarily invented integral and differential calculus to help solve the problem.

Group theory, the mathematical description for families of properties, has widened our understanding in many disciplines that range from solutions of the Rubick Cube to elementary particle physics.

There are also trigonometry, differential geometry, number theory, complex analysis, vector calculus, topology, boolean algebra, and so on. Some have proved to be indispensable to the advances of the physical sciences, while others remain mathematical curiosities.

Mathematics is the purest product of logical thinking—a creation of the human mind. It is often marveled that mathematics can describe nature at all. Scientific research benefits greatly from this happy fact. Physical models and theories become remarkably useful when given a mathematical translation because it ensures that only logical conclusions are drawn from the original assumptions. It also provides a vehicle to help one contemplate abstract concepts. Often it is the manipulation of an equation that indicates a physical prediction or property that may not otherwise be high-lighted. For example, our entire understanding of classical gravity can be derived from a single, relatively simple formula laid down by Isaac Newton. And the equivalence of energy and mass, a concept that irreversibly shaped twenti-eth-century physics and wartime politics, is described by Einstein's famous formula $E = mc^2$, which is discussed further in chapter 6.

You may have been taught in eighth grade algebra that not only does three times three equal nine but so does nega-tive three times negative three; in other words, the square root of nine has two equally valid solutions. In the late 1920s there was a mathematical discovery that had scientific as well as literary importance: the theoretical physicist Paul Dirac explored a second (negative) solution to his equation for an electron. From this solution Dirac boldly predicted the existence of antimatter—that favorite fuel of science fiction writers. And in 1932 the American physicist Carl Anderson discovered the positron, the antimatter partner to the electron.

All scientists are familiar with some level of higher mathe-matics, and while the world's nations speak different lan-guages, everybody's mathematics looks the same.

The Right Question

Wind is caused by the trees waving their branches
—Ogden Nash

You will agree, no doubt, that it is possible for a question to be grammatically correct (with subjects and verbs in the right places) yet at the same time be physically meaningless or make no sense at all. Some examples:

At what temperature does the number seven melt?
What is the sound of one hand clapping?
What is the square root of a pork chop?

In the sciences, it is also true that your query is not guaranteed to have meaning or significance just because you assemble words in the form of a question.

Perhaps half the effort to answer a question occurs when the question itself gets formulated. In general, the fewer assumptions that are contained in a question, the more tractable is the solution. A poorly formed question might read, "How far is the edge of the universe from the center of the universe?" The question, as constructed, assumes that (1) an edge of the universe exists, (2) a center of the universe exists, and (3) there exists a method by which this distance can be measured. Clearly, if the universe has no edge then the entire question has no meaning. Ideally we would use several questions:

> Does the universe have an edge?
> Does the universe have a center?
> How can this distance be measured?
> What is this distance?

The original question is now subdivided into morsels of concepts that may now be addressed individually, and in sequence.

The most successful of all scientists are those who consistently ask the right question. The mind-scrambling inverse of this talent was displayed by the researchers of the Lagado Academy in Jonathan Swift's 1726 satire, *Gulliver's Travels*. They are unforgettable examples of people who conduct meaningless research because they are not guided by the right questions. In one such instance, a Lagado astronomer places a sundial atop a weather vane of the local townhouse. By adjusting the sundial's reading for Earth's rotation and revolution around the Sun, he attempts to deduce which way the wind blows. Then there is the Lagado chemist who labors to turn marble into soft pillows. In another example, a Lagado inventor, in an attempt to design a cheap method to plow the soil, recommends that farmers bury rows of tasty grubs for hogs to dig up and eat. After a few days, the hogs will have dug up the whole ground in search of their food and thus make it fit for sowing, while at the same time "manuring it with their dung." The fact that one would have to invest great effort to dig up the soil and bury the food in the first place seems to elude the inventor.

As scientists, if we do not suitably guide the questions we ask, then we may also fall victim to asking questions that transcend the available data or even transcend the realm of experiment. Consider the following: "What existed before the universe?," "Why do energy and matter exist?," "Is our universe a single atom of a larger megauniverse?," and "Can the human mind ever understand the universe if the human mind is part of the universe?" These are all—at least for now—questions of philosophy and metaphysics.

What's in a Theory?

The great tragedy of Science—the slaying of a beautiful hypothesis by an ugly fact.
—Thomas Henry Huxley

The power of a theory lies not with what it predicts that is already known, but with what it predicts that remains to be discovered. An excellent scientific theory is one that is not only based on data but is falsifiable, predicts new phenomena, and unifies previously disjointed sets of ideas. Several of the most successful theories and discoveries in all of science are listed below:

- · Newton's laws of gravity
- · Thermodynamics
- · Darwin's origin of species
- · Mendeleev's Periodic Table of the Elements
- · Hertzsprung-Russell (H-R) Diagram
- · Einstein's theory of relativity
- · Quantum mechanics
- · Watson and Crick's deoxyribonucleic acid (DNA)

There exists, of course, a much longer list of theories that have fallen to the junkyard of failed ideas.

After the dawn of the twentieth century, scientists stopped labeling successful theories as "laws." What helped to instigate this change of vocabulary was that twentieth-century physics opened new experimental domains where the predictions of previous laws were inadequate. It was the humble recognition that newer and better equipment may provide data that will indicate a deeper realization of the physical world. This is why before 1900 we had Kepler's *laws* of planetary motion, Newton's *laws* of gravity, and the *laws* of thermodynamics whereas after 1900 we have Einstein's *theory* of relativity, quantum *theory*, big bang *theory*, and so forth. It is important to remember that the universe doesn't care much at all about our laws and theories. It doesn't obey our laws any more than the Sun obeys us when we say, "The Sun will rise at 6:58 tomorrow morning."

Some theories are more tested than others. A good example of a theory in progress is the big bang description of the origin of the universe. It is consistent with most of the available data, which makes it the best supported of all theories of the universe. If the big bang theory continues to be consistent with new data, then we may place more confidence in the concept. Eventually, if its success continues, there will be a day when conflicting data will be held suspect unless the evidence is truly compelling and irrefutable. Only then will we be forced to rethink the theoretical framework of the universe's origin. This is why data that conflict with a well-established and well-tested theory must be interpreted with extreme skepticism. To alter established physical theory simply to explain somebody's peculiar data ought to be considered something beyond a last resort. Such claims of discoveries often attract media attention. The public needs to remember, however, that resistance to ideas that overthrow a scientific worldview is a productive and healthy posture. In the research community, we are simply being skeptical and cautious, not self-righteous and stubborn. History has shown that the overwhelming majority of ideas that conflict with established physical law turn out to be wrong. These failures usually pass undocumented by the public press.

Many people who are not formally trained in mathematics or physics want to be "armchair" Einsteins. I have a file drawer that bursts with people's pet theories on the evolution of the universe. It is not often appreciated that major theoretical leaps of science typically come from those who know nearly everything that has come before them. When combined with a spark of scientific inspiration, a successful theorist then sees what everyone else sees, yet thinks what no one else has thought. This does not normally happen from an armchair tyro in the absence of extensive training—especially since one may not even know what questions to ask.

Some theories are scientifically empty. An early example includes the following line of argument that is traceable to Aristotle and was adopted by the Roman Catholic church through to the seventeenth century: God created the universe; God is perfect, so must be the universe; the circle is a perfect geometric shape; therefore all orbital motion must be perfect circles. A similar argument was propounded to defend why Earth must be in the center of the universe: God created the universe; God created humans on Earth in His own image, humans must be special; the center of the universe is special; therefore the Earth is in the center of the universe. Well, all orbits we have ever measured are flattened circles, and we have not found Earth to be in the center of anything. Theories without physical basis are fragile indeed.

Some types of theories are of no utility to science because they make no predictions, or the predictions are untestable. For example, "At the beginning of the universe, seven separate universes were formed. Ours is one of them. The other six are outside our universe so they cannot be detected." These are the kinds of theories you should keep to yourself. Other theories conveniently explain all that is known in ad hoc fashion—a separate and exotic cause is invented for each phenomenon—yet they do not lend insight to new discoveries. Erich von Däniken, author of the bestseller *Chariots of the Gods*, was notorious for this approach. He continually explained ancient monuments and stone carvings that he did not understand by attributing their origin to extraterrestrials that visited Earth eight to ten thousand years ago. Before Sir Isaac Newton published the *Principia*, his 1687 treatise on gravitation, others had hypothesized that there were different mechanisms that kept the moon in orbit, rocks on the ground, and projectiles in their ballistic paths. The English scholar William Occam suggested in 1340 that

Multiplicity ought not be posited without necessity.

In regular English, Occam meant that the simplest explanation with the fewest assumptions is likely to be the correct explanation for a phenomenon—a principle now called "Occam's Razor." Newton published a similar comment:

> We are to admit no more causes of natural things than
> such as are both true and sufficient to explain their appear-
> ances. To this purpose . . . Nature is pleased with simplic-
> ity and affects not the pomp of superfluous causes.

Guided by these principles and his towering genius, Newton
weaved all of gravity into one concept and one fundamental
equation. Guided by the opposite of these principles—where
the most extraordinary explanation is preferred—Erich von
Däniken sold millions of books.

Some other theories are constructed so that they are not
falsifiable. Herein lays the primary schism between deity-
based mythology and scientific theory. If all that you see,
do, measure, and discover is the will of a deity, then ideas
can never be proven wrong, you have no predictive power,
and you are at a loss to understand the principles behind
most of the fundamental interconnections of nature.

Ongoing science is not without conflict. Occasionally the
general public, by way of news reporters, gets to eavesdrop
on scientific discourse. If the topic is "hot" then the public
will see several (or more) scientists in active debate about
some theory or hypothesis. The lack of agreement is a normal, healthy part of the progress of science. Usually the
disagreements are resolved when appropriate data become
available, but often it is the exchange of ideas that serves to
guide how the resolving data are to be obtained. The exchange of ideas occurs at all conventions, all conferences,
and in the hallways and coffee lounges of research institu-
tions.

The prospect of a profound discovery serves as an important research incentive, but it is the process of science
that excites the scientist. It occurs daily for most research
scientists and continues nightly for many astronomers. And
like the busy construction of an anthill—where each ant

carries a single grain of sand with a single purpose—we design and build experiments, develop theories, generate computer simulations, invent new methods of analysis, solve problems, and publish the results for the international community of scientists to see. These are the activities that are *along* the yellow brick road. One's formal training, when combined with insight, and sometimes luck, is what improves the chance that you will find yourself on the correct path to discovery and thus contribute to the growth of this great tapestry we call science.

· 3 ·

Measure for Pleasure

No science attains maturity until it acquires methods of measurement.

—*Logan Clendening*

☙

*P*erhaps the first fraction anybody learns is *one-half*. From about age five onward this is what you likely used to append to your age when asked how old you were. Three weeks after your fifth birthday you probably proclaimed, "I'm five and a half years old." You were older than five and younger than six, and more complicated fractions were yet to be learned. Little did you know that all measurement that has ever occurred suffers a lack of precision not unlike your early age estimate.

If you were a precocious child who knew there are fifty-two weeks in a year, you might have proclaimed, "I am five and three fifty-secondths years old." An even more precocious child might have gone on to inquire whether the question was asked exactly twenty-one days after the fifth anniversary of the moment of your birth. If this was indeed the case, then someone else (perhaps a five-year-old child who is ready for college) might indicate that you took four seconds to declare "I am five and three fifty-secondths" and then want to know which of the four seconds occurred at exactly five years and three weeks.

Such a conversation could of course continue without limit: through tenths of a second, hundredths of a second, thousandths of a second, and so forth. The circumstances are further confused because mammalian births do not happen instantaneously, and all calendar years are longer than fifty-two weeks.

Adults are as guilty as children when it comes to the fraction *one-half*. Have you ever heard someone proclaim, "I am 5 feet 7 and ³/₈ths inches tall"? Not likely. People tend to round up in units of one-half inch. But even if they didn't, one's height cannot be measured more precisely than the smallest unit of division on the measuring stick. In America, that unit tends to be one-sixteenth of an inch.

Your measured height is likely to fall between two tick marks, which might prompt you to engrave even finer divisions on the measuring stick. Eventually, you may find that your height falls exactly on a tick mark—perhaps the 5 feet 7 and $^{45}/_{128}$ths-inch tick mark. But the grim truth is that the tick mark itself has a thickness within which your true height falls. It should be clear by now that no matter how finely spaced the tick marks are placed on a measuring stick your height cannot be determined exactly—only more and more precisely.

However, if you are six feet tall and of the human species, then there was a time in your life between when you were born and now when you were *exactly* 5 feet 7 and $^{45}/_{128}$ths inches tall. This remains true in spite of your inability to measure it as such to unlimited precision. Age and height are only two of many quantities that vary continuously and cannot be measured exactly. Other common examples are temperature, volume, and weight. Note, however, that you cannot simply add temperatures the way you can add volumes or weights. Two people in bed, each with body temperatures of 98.6 degrees Fahrenheit, do not normally create a 197.2 degree under-the-cover oven.

If all this talk about imprecision upsets you, then you may be happy to know that discrete entities can be reckoned exactly. You (probably) have exactly one head. A rectangle has exactly four corners. You can cut a pizza into exactly eight slices. Here, of course, we are simply counting rather than measuring.

Some physical entities involve more than one measurable unit. Speed is always given as some unit of distance "per" some unit of time. The "55" speed limit that is occasionally obeyed on the nation's freeways implies the units of miles per hour (mph). You can know your speed only as accurately as you can measure the length of a mile and the duration of an hour. In practice, however, your accuracy is thwarted by sources of much greater uncertainty such as low-quality speedometers, tire size, and tread depth.

Apart from inchworms, residents of the United States, and the Aborigines of the Australian Outback, the metric system is the basis of nearly all measurement by the Earth's population. The metric system was first proposed in 1670 by Gabriel Mouton, the Vicar of Lyons, France, but it was not in widespread use until the French Revolution of 1789 provided the political and social vehicle to disseminate it. By 1795 the metric system was formally adopted throughout France. This new system of weights and measures then spread rapidly throughout the civilized world.

The American fear of the metric system has very little basis. True, no one will deny that many units of the metric system sound like an ensemble of insects that crawl in your garden (millimeters, centigrams, kilometers, etc.). But nobody will force you to say, "A gram of prevention is worth a kilogram of cure," and you can even continue to use words such as *mile*stone and *pint*-sized. In general, the metric system is defined through physical concepts rather than through correspondence with everyday things such as cups, somebody's feet, horses, candles, and bushels. But I am worried: since the American system elects to measure car engines in "horse power," then will American intergalactic spaceships be measured in "car power"?

Behind the scenes of ordinary measurement are the metrologists—scientists who work to define the basis of the metric system itself. At the 1967 General Conference on Weights and Measures, the *meter* was defined as the length of a designated number of cycles of the orange-red photons that the atom krypton-86 can emit when it is excited. The *second* was defined as the duration of a designated number of periods of the photon of light that corresponds to a specified atomic transition within cesium-133. But how well can we measure the wavelength of the orange-red light from krypton-86? How reliable is the photon's period that is associated with cesium-133? These uncertainties set the ultimate experimental limits on the measurement of length and time.

The first attempt to measure the speed of light was by Galileo in the early 1600s. He sent an assistant to a distant hill to flash the light of a lantern. Galileo responded immediately with flashes from a lantern of his own. An attempt to time the delay proved futile—human reflexes were inadequate for such a task. Galileo succinctly noted that light "if not instantaneous . . . is extraordinarily rapid." One could call this a low-precision estimate.

In 1675 the Danish astronomer Ole Röemer noticed that, as seen from Earth, eclipses of Jupiter's moons occurred systematically earlier and later than expected. Röemer was clever enough to realize that the eclipses occurred early when Earth, in its orbit, was nearest Jupiter, and they occurred late when Earth was farthest from Jupiter. Röemer deduced correctly that these time differences were due to the time it took for light to span the diameter of Earth's orbit. By simple division (diameter of Earth orbit ÷ total time difference = speed of light), Röemer provided the first reasonable estimate for the speed of light.

Albert Einstein later postulated through his theory of relativity that in a vacuum, no matter your state of motion, you will always measure the speed of light to be the same—within experimental accuracy. Such a theory, coupled with continued and irrefutable experimental support, has endowed the speed of light with an almost sacred importance. Over three hundred and fifty years after Galileo's first estimate, the precision in the speed of light has finally reached the experimental limits imposed by the meter and the second. This limit corresponds to an accuracy of about one part in ten billion.

To be fair to the cesium atom, the second can be measured to about one part in one hundred trillion, so our constraints on the precision of the speed of light can be blamed on the meter. In other words, if we knew the length of the meter as precisely as we know the duration of a second, then we could know the speed of light ten thousand times more accurately. This frustration has led the International Committee on Data for Science and Technology, in an unprecedented decision, to *define* the speed of light at the current best value: 299,792,458 meters per second. For the future, improved precision in the speed of light will translate directly to a modification in the length of the meter so that the defined value for the speed of light is preserved. The definition of the meter is now: the distance traveled by a beam of light in a vacuum during 1/299792458th of a second.

Future modifications to the length of the meter will not be of the kind that ordinary people need to worry about. You will not wake up one morning to find that you suddenly measure twice as tall as when you went to sleep. Meter modification would occur on a scale that is smaller than the thickness of the molecules that compose the measuring edge of a household ruler. That being said, the quantum behavior of the physical universe on the smallest of scales ensures that space itself is discontinuous, which leads to an ultimate precision with which the length of the meter can be defined. As was noted in chapter 1, you cannot measure precision, or lengths in general, on scales smaller than the teeny-weeny "Planck length" of 1.6×10^{-33} centimeters. Anything smaller is physically meaningless.

Digital clocks have given society a false sense of accuracy.

Several years ago I was one of the tens of thousands of people that assembled in New York City's Times Square to celebrate New Year's Eve. While there, I took note of three different digital clocks that were part of large neon billboards. Two of them displayed time to the tenth of a second. The other displayed time to the hundredth of a second. A fourth digital clock high on a building in the middle of Times Square was the clock that most people watched. At precisely midnight, whenever that was, *none* of the clocks agreed. A full ten seconds separated the fastest from the slowest clock. At least three clocks—and quite possibly all four—were wrong with impressive precision. Most people assembled for the occasion took no notice, but I was devastated. As the adage goes, "The person with one clock knows the time. The person with two clocks isn't sure."

Hand-held calculators and computers in general have also given society a false sense of accuracy. For example, if you drive a car at 23 mph and your destination is seven miles away, and if you are one of those people who never leaves home without a calculator, then you might take pleasure in computing how much time your trip requires. So you divide 7 (miles) by 23 (miles per hour), and your calculator promptly displays 0.3043478261 (hours), which of course comes to 18 minutes and 15.65217391 seconds. Nobody measures the duration of car trips to the billionth of a second— but your calculator doesn't know this. It also doesn't consider that there will be traffic lights and stop signs and pedestrians that normally require you to stop once or twice, thus rendering the precision of the calculator computation completely useless. It is up to you to say, "We will arrive in about twenty minutes."

The airline industry is also guilty. Moments after takeoff the pilot announces your trip will take 2 hours and 47 minutes. What the pilot should say is, "My on-board computer tells me we will be airborne for 2 hours and 47 minutes, but all I can say with assurance is that we will land within several minutes of the computed time."

The entire (responsible) scientific community, when reporting a measurement or computation, will give the reader an indication of the believability of the results. A convenient symbol, which often precedes an uncertain quantity, is the single or double tilde: $\sim$ or $\approx$. They are the scientific equivalent of an "–ish" as in, "Dinner will be served around six-ish". Better yet, if the range in error estimate is known, it may be noted by the famous "plus-or-minus" symbol which is simply a plus sign sitting atop a minus sign: "$\pm$". It is more than a math symbol—it is a symbol of honest uncertainty. The next time somebody tells you an exact-sounding quantity just respond, "Plus or minus what?"

In summary, then, we can say:

Our precocious preschooler was 5 years 6 months $\pm$ 6 months old.
Your height is 5 feet 7 inches $\pm$ ½ inch.
The speed of light is 299,792,458 $\pm$ 0 meters per second.
The car trip of the calculator addict will last 20 $\pm$ 5 minutes.
Your plane trip will last 2 hours 45 minutes $\pm$ 15 minutes.
And, of course, you have exactly one head.

· 4 ·

The Confused Person's Guide to Astronomical Jargon

This chapter might possibly belong at the end of the book as a glossary, but a discussion of the methods of science would be incomplete without paying homage to the invention, development, and usage of jargon. An academic discipline that is sufficiently mature will have normally assembled for itself a jargon-filled lexicon. But before you get indignant about this, consider that academic researchers are not exclusively guilty. When was the last time you understood your car mechanic when you were duly informed of what was wrong with your car? And if baseball were not your passion, then the following plausible scenario would sound completely meaningless: "The DH, who had homered in each of his first two at-bats, reached first on a pitcher's balk, and two outs later advanced to third on a ground-rule double. He then scored to win the game on a payoff pitch to a batter who laid down a bunt for the squeeze-play in the bottom of the tenth." Yes, we all have our jargon, and we all use it to communicate with others in our field. But in my (possibly biased) opinion, astronomy has the most entertaining jargon of any discipline—enough to warrant a chapter of its own.

☙

Terms of Entearment

To a botanist, the North American rose is a *Rosa nutkana*. To a marine biologist, a household goldfish is a *Carassius auratus*. To a medical doctor, a bruise on your jaw is a mandibular contusion. To a sociologist, your next-door neighbor is your residential propinquitist. These professions, and many others, are replete with polysyllabic terms that are precise yet devoid of romance. To Juliet Capulet (of *Romeo and Juliet* fame), "a rose by any other name would smell as sweet." But what Juliet neglected to mention is that a rose by a five-syllable term would make its way into much less poetry.

Astronomers, however, get the award for creating the most diverse set of terms ever assembled to communicate science. There are romantic-sounding words, words that mean something different from what they say, words that are intentionally misspelled, words that sound like diseases, words that are historical relics, and most importantly, household words that mean exactly what they say. Consequently, terms of astronomy can be enlightening as well as mind-scrambling—but never boring.

Some Terms That Mean Exactly What They Say

Red Giants: This is what we call big red stars. It is an evolutionary phase through which nearly all stars pass.

White Dwarfs: This is what we call little white dead stars. Only rarely is "dwarfs" spelled as "dwarves." Do not confuse these with dwarf stars, which are "living" main sequence stars such as the Sun that burn hydrogen for fuel in their core.

Black Holes: This is what we call gravitational holes in space and time that look black. A black hole's surface gravity is so strong that the speed one needs to escape from them is greater than the speed of light. Since light itself cannot escape, which is why black holes look black, then all hope would be lost for you if you happened to stumble upon one. Unlike a simple hole in the floor, you can fall into a black hole from any direction. Yes, the properties of black holes would make good script material for a science fiction horror story.

Big Bang: This is the technical term we use to describe the beginning of the universe. It must have been a really big explosion, even if nobody was around to watch or listen. It is estimated to have occurred about 15 gigayears (15×10^9 years) ago.

Missing Mass: This refers to dark matter in the universe that we have good reason to believe ought to be out there, but we cannot see it. We are still looking for it.

Star Cluster: This is a cluster of stars that are held together by their collective gravity. One variety of cluster contains relatively few (up to a thousand) stars and has an open appearance. We call these *open clusters.* Another variety is globular in appearance and can contain up to hundreds of thousands of stars. We call these *globular clusters.*

Star Formation: This is the official term to use when it is time to discuss the formation of stars.

Some Famous Names That Precisely Describe the Object's Appearance

Jupiter's Red Spot: There is a large circular red region on Jupiter's cloudy surface. It is a raging anticyclone several times larger than Earth that was discovered by Galileo more than 350 years ago. It is officially called the *Red Spot.*

Incidentally, the planet Neptune has a big (dark) spot of its own, which is officially called Neptune's *Dark Spot.*

Sunspots: On the Sun's visible gaseous surface are small areas that are cooler than the surrounding regions. Relative to the rest of the Sun, these spots look dark. Ignoring the fact that they are periodic magnetic storms that move in pairs across the Sun's disk, we simply call them *sunspots.*

Ring Nebula: The tenuous outer envelope of what was formerly a red giant star has escaped into interstellar space. It is nebulous and it looks like a ring. We call it the *Ring Nebula.*

Crab Nebula: Even though there are no claws, no roaming eyeballs, and no antennae, this nebulous explosive remnant of the famous supernova of A.D. 1054 resembles what an impressionist artist might draw as a crab.

Horsehead Nebula: In a corner of the constellation Cygnus there is a dark cloud that obscures part of an illuminated gaseous region behind it. The dark cloud bears a remarkable resemblance to the silhouette of a horse's head.

Milky Way: If thoroughfares of ancient times had been called "streets" instead of "ways," then our galaxy might have been named *Milky Street.* Without a telescope, the billions of stars that compose our galaxy are distant enough, and dim enough, to blend together in what resembles a milky path across the sky. The milk theme also exists in the word "galaxy" itself—the Greeks called the Milky Way the *galaxias kuklos,* which translates to "milky circle."

Some Terms That Sound Mysterious

Albedo: Pronounced "al-bee-dough," it is simply a measure of how much light a surface reflects. A perfectly white surface will reflect all light and have an albedo of exactly

1.0, while a perfectly black surface will absorb all light and have an albedo of 0.0.

Zone of Avoidance: The solar system is embedded in the star-filled, gas-rich, and dusty disk of our Milky Way galaxy. We must look above and below this galactic pancake to see other galaxies and the rest of the universe. A map of all objects in the sky will readily show that galaxies seem to "avoid" this zone where our own galaxy's disk is in the way.

Event Horizon: The boundary between what is in our universe and what is not in our universe. For example, it is the horizon that separates us from the undetectable galaxies that recede with the speed of light at the "edge" of the universe. Additionally, the event horizon of a black hole is what separates us from the region where light (and anything else) cannot escape. Indeed, the size of black holes and the size of the universe are defined by their event horizons.

Roche Lobe: In the mid-nineteenth century, the astronomer E. Roche studied the detailed gravitational field in the vicinity of a binary system. The Roche lobe is an imaginary, dumbbell-shaped, bulbous envelope that surrounds any two orbiting objects. What makes the Roche lobe special is that if material from one object passes across its own envelope, then the material is no longer gravitationally bound. This peculiar-sounding event is actually common among binary stars where one star swells to become a red giant as it overfills its Roche lobe. The material then spirals toward the second star, which adds to its mass, thereby hastening its evolution. When the second star becomes a red giant, the mass-transfer will reverse, thus creating a modeling nightmare for binary star theorists.

Some Terms That Sound Like Names One Might Give to an Alien

Perigalacton: For anything in orbit around a galaxy (inclusive of another galaxy), it is the point of closest approach. The farthest orbital point is, of course, *apogalacton.*

Boson: What at first sounds like the name given to residents of planet "Boso," is actually the collective name given to particles with a specific quantum mechanical property in common. This includes all *photons* (massless particles of light) and all *mesons* (elementary particles with masses that fall between that of the electron and proton). Bosons are named for the Indian physicist Jagadis Chandra Bose.

Baryon: Another group of particles, named from the Greek *barus,* meaning "heavy." These are neutrons and protons and all heavier particles that decay to become them.

Omega Centauri: Surely there must have been a person, place, or thing on the *Star Trek* television and film series that was an "Omega Centauri." In astronomy, however, it is the name given to the titanic globular cluster of stars that appears in the southern constellation Centaurus.

Some Terms That Look Like Typographical Errors

Gnomon: A vertical stick in the ground (*not* the otherwise well-known stick-in-the-mud) that was used by the ancients to measure the angle of the Sun above the horizon. By knowing the height of the stick and by measuring the length of its shadow, one can determine the altitude of the Sun with great precision. The same term is used for the raised pointer of a sundial. A gnomon is useless on a cloudy day.

Analemma: If you place marks on the ground at the top of your gnomon's shadow at exactly the same time of day, for every day of the year, then the pattern of marks will trace a figure "8." This is a simple demonstration that the Sun does not always return to the same spot in the sky at the same time each day. The figure "8" is called an ana-lemma and is often inscribed in sundials or drawn on globes of the Earth—usually somewhere in the Pacific Ocean.

Syzygy: A favorite of crossword puzzle enthusiasts, this less-than-elegant term describes the moment when three cosmic bodies have aligned. For example, during full moon and new moon, the Earth, Moon, and Sun are in syzygy.

Gegenschein: A faint glow seen in the nighttime sky 180 de-grees away from the sun. It is the reflection of sunlight back to Earth from particles in the plane of the solar sys-tem. Gegenschein translates from the German as simply "reflection."

Ylem: The American physicist George Gamow suggested this name for the high-temperature primordial cosmic soup that preceded the big bang. George Gamow is no longer with us, and neither is his word.

Orrery: Any mechanical model of the solar system where planets can actually revolve around the Sun. The better models also display the various moons that revolve around the planets.

Some Terms That Don't Carry Emotional or Intellectual Stigma

Mean Sun: Here "mean" means "average." Because of Earth's elliptical orbit, and because the Sun does not tra-verse the sky along the celestial equator, the Sun does not always take 24 hours to reach its highest point in the daytime sky. Sometimes it takes less, sometimes it takes more. Also, atmospheric refraction makes the Sun appear to move through the sky more slowly than it otherwise would. To render the Sun more friendly to timekeepers, we define the average Sun as simply the one that moves uniformly through the sky so that it always takes 24 hours to reach its highest point. All the clocks of society are set to the mean Sun and grouped, for convenience, into time zones.

Major Planet: Any of the nine known planets (i.e., Mercury, Venus, Earth, Mars, Jupiter, Saturn, Uranus, Neptune, and Pluto).

Minor Planet: Any asteroid.

Inferior Planet: Refers to the planets found between Earth and the Sun (i.e., Mercury and Venus).

Superior Planet: This refers to the planets found beyond Earth's orbit (i.e., the rest of them).

Unstable isotope: One of the few properties of atomic nuclei that is modified by a word also used by psychopathologists. The identity of a chemical element is set by the number of protons in its nucleus. The number of neutrons, however, can vary. Each variation in the number of neutrons is called an isotope of that element. But deep in the world of atomic nuclei, life is not always tranquil. Some elements have isotopes that are decidedly unsatisfied with the number of neutrons they contain. These elements can transmutate (decay) into another element by converting one of their neutrons into an electron and proton pair. Such disgruntled elements are quite sensibly referred to as unstable isotopes.

Great Circle: This is, quite simply, the shortest distance between two points on the surface of a sphere. There is nothing especially great about it, and it normally refers to only a segment of a circle, such as the path an airplane might take between two cities.

Eccentricity: The mathematical measure of the shape of a simple orbit. For example, a perfect circle has zero eccentricity. An ellipse can have an eccentricity anywhere between zero and one. A parabola has eccentricity equal to one, and a hyperbola can have any eccentricity greater than one. Eccentric orbits are easier to understand than eccentric people, but perhaps slightly less interesting.

Degenerate Star: Any star that is supported from collapse by an interparticle pressure that prevents electrons from getting too close to one another. This *electron degeneracy* supports all white dwarfs from further collapse. Neutrons can also support a star from collapse by a similar mechanism. This *neutron degeneracy* supports all neutron stars (inclusive of pulsars). White dwarfs and neutron stars contain some of the densest known matter in the universe.

Greatest Brilliancy: Used almost exclusively for the planet Venus, the term refers to the moment when the planet, in its orbit around the Sun, is brightest as viewed from Earth. When this happens, Venus is likely to be near the horizon and brighter than any airplane in the sky. Since Venus is not "coming in for a landing," people who do not know better tend to call police departments to report Venus as a glowing and hovering UFO.

Some Terms That Sound Like Diseases

Inferior Conjunction: Refers to when an inferior planet passes between Earth and the Sun. This is not a very interesting event because it is the other side of the planet that is illuminated with sunlight.

Superior Conjunction: Refers to when a planet (inferior or superior) passes to the other side of the Sun from Earth. If for no other reason, inferior and superior conjunctions are worth noting because they are the moments in a planet's orbit that signal the transition between when a planet is visible in the evening sky and morning sky. For both inferior and superior conjunctions, the Sun, Earth and the planet are in syzygy.

Occultation: Refers to when a foreground object passes in front of a background object. The term is typically used when an asteroid or the Moon passes in front of either a planet or a star. Strictly speaking, a solar eclipse is an occultation.

Obliquity of the Ecliptic: This is simply the tilt angle of Earth's axis. More complicatedly put: when the plane of the solar system (the ecliptic) and Earth's equator are each projected onto the sky, they will intersect at an angle called the obliquity of the ecliptic.

Bok Globules: What sounds like malignant tumors is actually the name for small, opaque regions of gas and dust in the Milky Way galaxy that are the sites of forthcoming star formation. They are named for the pioneering Dutch astronomer Bart Bok.

Some Terms That Sound Like They Could Be Rock Groups

Shadow Bands: These are fleeting ripples of shadows that are noticed during a solar eclipse just before and just following totality. With the Sun as a skinny crescent, the atmospheric optics are ideal for revealing the fluctuations in density within Earth's lower atmosphere.

Kirkwood Gaps: Regions in the asteroid belt between the planets Mars and Jupiter where orbits are unstable and almost no asteroids are found. Named for the nineteenth-century American astronomer Daniel Kirkwood, who first explained the effect. A gap with similar dynamical origin is found in the rings of Saturn, except that it is called

Cassini's division, after the seventeenth-century Italian astronomer Giovanni Domenico Cassini.

G-Band: A region of a star's spectrum that has a strong absorption feature from the presence of the two-atom molecule carbon hydride (CH) in the star's atmosphere.

Atmospheric Band: A region of a star's spectrum that has a strong absorption feature from the presence of the oxygen molecule (O_2) in Earth's atmosphere. All starlight that is observed from Earth's surface must pass through the atmosphere. Consequently, this band sneaks into the spectrum of every star.

Some Terms That Have Too Many Syllables

Magnetohydrodynamics: In the Germanic tradition of slapping together word parts to make an even bigger word, magnetohydrodynamics is the study of the effects of a magnetic field (*magneto*) on the behavior and motion (*dynamics*) of a fluid (*hydro*) that is hot enough for electrons to be separated from their host atoms. Such a gas is called a plasma and is often considered to be the fourth state of matter.

Thermonuclear Fusion: It takes very high temperatures (*thermo*) to merge (*fusion*) positively charged atomic nuclei (*nuclear*) against their natural force of repulsion to create heavier atomic nuclei. The core of the Sun merges hydrogen atoms to form the heavier helium atom with an enormous energy dividend. This nuclear reaction powers the Sun and, in a less contained way, also powers H-bombs.

Spectroheliograph: A device that solar astronomers use to observe the Sun in a narrow part of the spectrum. Often the intent is to isolate a single emission or absorption feature.

Orthoscopic Ocular: One of many different varieties of telescope eyepieces, this one is relatively expensive and is good if you want excellent image quality. It is also one of the few eyepieces that work well with bespectacled observers.

Some Terms That Sound Like Romantic Places

Coudé Room: It sounds romantic as long as the English translation of the French *coudé* is not revealed. Some telescopes have secondary and tertiary mirrors whose adjustable placement can considerably extend the path of starlight before it comes to what is then called the coudé focus. En route to the detector, the starlight is directed out of a hole in the telescope's side and focuses in a separate room—the coudé room—where high-resolution spectra are recorded. (And in case you are still wondering, the English translation of *coudé* is "elbow.")

Ascending Node: The spot in space where a tilted orbit crosses a preestablished plane going north. When the orbit crosses the plane going south the node is descending. It is commonly used when planet orbits and binary stars are discussed.

Lagrangian Point: What sounds like it ought to be the name of one of those erogenous spots on the human anatomy is actually any one of five points in the vicinity of two orbiting bodies where all centrifugal and gravitational forces balance. It is named for the eighteenth-century French mathematician J. L. Lagrange. One of the five Lagrangian points, "L-5," had been adopted as the original name of a space exploration society that seeks to promote, among other things, the construction of a space station at this location of the Earth-Moon system.

Some Terms That Are Historical Relics

Spectral Lines: In the old days of astronomy, when photography was the standard means of detection, it was common to publish photographs of stellar spectra. A typical stellar spectrum produced by a prism or a diffraction grating will display an elongated rectangle of light that is marked with narrow emission and absorption features, which indicates (among other things) the chemical composition and temperature of the star's atmosphere. In a photograph, these features look like lines that segment the rectangle—hence the term *spectral lines*. Nowadays, with modern digital detectors, spectra are commonly published as graphs of intensity versus wavelength (or something equivalent to wavelength). In these displays the word *line* loses its descriptive meaning. The emission features look like peaks, and the absorption features look like crevasses—but they are still called "lines." It is not a singular tragedy, however. There are plenty of examples where word meanings have changed due to technology. For example, some people still call the refrigerator an "icebox," and many people still say they "dial" a telephone number even though they are simply pushing buttons.

Redshift: This word can lead to confusion if taken literally. Formerly two separate words, it soon became hyphenated. In its current use, "redshift" has finally lost its hyphen. When used among astronomers who study galaxies, it refers to the shift in spectral features (absorption or emission lines) toward the longer wavelengths that is the consequence of a galaxy's motion away from us. In an expanding universe, where distant galaxies recede faster than nearby galaxies, the redshift is frequently taken to be an indicator of distance. All the spectral lines that were first used to measure this shift had shorter wavelengths than red light. A measured "red shift" therefore had the unambiguous meaning that the spectral features shifted toward the longer wavelengths of the red part of the spectrum. Yet infrared, microwaves, and radio waves all have longer wavelengths than red light. A feature in these parts of the spectrum, if it experienced what is called a redshift, would still shift to longer wavelengths—but to do so moves the feature *away* from the red part of the spectrum, not toward it. A less confusing (though nonhistorical) name might be "longshift."

Some Famous Acronyms

Laser: Laser, like "scuba" and "radar," is one of those acronyms that has achieved greater status than the words for which the letters stand. Certain atoms and molecules, when excited, can be made to emit photons of visible light upon being stimulated by photons of the same energy. The remarkable result is an amplified "coherent" pulse of photons, all with the same energy. With a specially designed cavity, this unusual property can be exploited to sustain a narrow beam of coherent photons. The process was dubbed "Light Amplification by Stimulated Emission Radiation" or *laser*, for short.

Maser: Identical to a laser except microwave light is emitted rather than visible light. The molecules OH (hydroxyl), H_2O (water), and SiO (silicon monoxide) have each been discovered to be a source of maser energy in gaseous regions of our galaxy.

Pulsar: Quite obviously, a pulsing star. Many types of stars pulse. The term "pulsar," however, is reserved for rapidly rotating neutron stars where their magnetic field axis is tilted from their axis of rotation. As the magnetic pole sweeps past our field of view, we detect pulses of radiation. Not all neutron stars have the favorable geometry to be called pulsars, yet all pulsars are neutron stars.

Quasar: A loose assembly of letters from the phrase "quasi-stellar radio source." With few exceptions, quasars look like ordinary stars on ordinary photographs. Their enormous redshifts and their staggering energy production make them some of the most curious objects in the sky. The first of these quasi-stellar objects to be discovered were strong radio sources. Later discoveries showed that most quasars are radio-weak. To be fair to this majority, the "radio source" was changed to "object" to now read "quasi-stellar object," or QSO, for short.

Some Terms That Have Nothing to Do with Punctuality

Early Galaxy / Late Galaxy: Early-type galaxies are elliptical, and late-type galaxies are open-pattern spirals. The original "tuning fork" galaxy classification diagram of Edwin Hubble displayed elliptical galaxies on a tuning fork's handle (extending to the left) with normal spiral galaxies placed along one tine, and spiral galaxies with a bar-pattern in their center placed along the other tine (each extending to the right). The spiral pattern became less tightly wound as you moved along the tines. Hubble postulated an evolutionary sequence among the galaxy shapes, but it was later found that no obvious connection exists. If you have difficulty remembering early from late, then imagine you are a snail on a page where the tuning fork diagram is drawn. If you started a left-to-right page trek, you would pass the elliptical galaxies *early* and the spiral galaxies *late.*

Early Stars / Late Stars: Early-type stars are hot, and late-type stars are cool. The original Hertzsprung-Russell (H-R) diagram plots luminosity versus temperature with the hotter part of the scale on the left and the cooler part of the scale on the right. Our page-trekking snail, moving once again from left to right (across an H-R diagram), will pass the hot stars *early* and the cool stars *late.*

Some Terms That Have Nothing to Do with Texture

Soft X-Rays: Low-energy X-rays. Nobody has ever squeezed them to verify that they are indeed soft.

Hard X-Rays: High-energy X-rays. Considerably more deadly than soft X-rays.

Some Terms That Have Nothing to Do with Distance

Near Infrared: If our peripatetic snail actually lived on the visual interval of a map of the electromagnetic spectrum (violet-indigo-blue-green-yellow-orange-red), and if our snail wanted to visit the infrared part that was just beyond the red, then it would consider the destination to be *near*.

Far Infrared: The snail would have to go *far* if it wanted to go beyond the near infrared to the part that was on the border with microwaves. Far infrared photons have much lower energies and longer wavelengths than near infrared and visual photons.

Some Terms That Have Nothing to Do with Etiquette

Proper Motion: This is the motion of a relatively nearby star when measured against the background of "fixed" stars.

Peculiar Velocity: For a star, this is the velocity that is left over after you have accounted for the larger-scale motion of the Milky Way galaxy's rotation. For a galaxy, it is the velocity that is left over after you have accounted for the larger-scale motion of the expanding universe. There is nothing particularly peculiar about either of these.

A Term That Has Nothing to Do with Jesus Christ

Right Ascension: As lines of longitude are used to locate east-west positions on Earth, so is right ascension used to locate positions of stars east-west in the sky.

Some Terms That Lie

Metals: Contrary to the tenets of a chemist, metals to an astronomer are all elements other than hydrogen and helium in the Periodic Table of the Elements. There is actually a practical utility to this scheme. The big bang endowed the universe with primarily hydrogen and helium. Everything else is "pollution" that was forged in the thermonuclear furnaces of stellar cores. Furthermore, in most environments that are astrophysically interesting (such as stars), the temperatures are so high that elements are vaporized and ionized into the free-floating charged particles of the stellar soup we call plasma. The traditional laboratory concept of a metal thus loses its meaning and significance.

Hydrogen Burning: This term is used by nearly all astronomers to describe energy production in the Sun's core. Conventional usage of the word "burn" refers to the breakup and rearrangement of molecular bonds with a release of chemical energy. But nothing actually burns in the Sun (not that all your possessions wouldn't vaporize if you tossed them there). It's just that the thermonuclear fusion of hydrogen in the Sun's core has no resemblance to any traditional understanding of the word "burn." Hydrogen fusion unleashes what is aptly called nuclear energy, which is not normally released in your household fireplace.

Planetary Nebulae: Everybody who owns a telescope and has looked skyward with it knows that stars do not look much different through a telescope than they do with the unaided eye—they just look like twinkling points of light. Planets, however, look like points of light only with the unaided eye. Through a telescope they become distinctive circular disks. The sky also contains fuzzy-looking things like galaxies, star clusters, and genuine gaseous nebulosities. One variety of nebulosity (the lost, over-puffed spherical envelope of a dead red giant star) often appears disk-like through a telescope. The visual resemblance to planets led to the unimaginative and misleading term "planetary nebulae."

Amateur Astronomer: If you put the word *amateur* in front of most professions, you would probably doubt whether a person with such credentials could be of any use to you. For example, it is not likely that an amateur neurosurgeon or an amateur attorney could attract much business. Amateur astronomers, however, are indispensable. Let it be known that the average amateur astronomer knows more about the appearance of the sky than the average professional astronomer. Furthermore, in almost all cases, the professional astronomer who knows the sky probably started as an amateur. The advantage to knowing what the sky looks like is that you also know when it looks different. Many supernovae, most comets, and nearly all asteroids are discovered by amateur astronomers upon noticing that a familiar region of the sky has a visitor.

Further Causes of Cosmic Confusion

What Do You Call Something That Is Big?

In the business of astronomy, if you dare call something big or bright you are at risk of exhausting your vocabulary of superlatives if you discover something even bigger or brighter. With open arms, astronomers have welcomed the word-prefix *super-* into the dictionary of cosmic jargon. It endows astronomers with the power to create terms like *supergiant, supercluster, superbubble,* and *supernova,* and it gives physicists terms such as *supercollider, supersymmetry, superstring, superconductivity, superfluid, supersonic,* and *superluminal.*

This penchant for using the word *super* has adequate precedent in twentieth-century society. Comic book characters with transhuman powers were always called superheroes. There are markets and supermarkets. There are highways and superhighways. There are ordinary bowls, and then there is the Super Bowl. The engines of some cars are charged while those of other cars are supercharged. And we can credit Walt Disney's Mary Poppins for the super version of "cali-fragil-istic-expi-ali-docious." A notable exception to this trend was the Boeing 747, which was spared being super in favor of the alliteration offered by *Jumbo Jet.*

In astronomy, giant stars are called giants. But when even bigger giants were discovered we were forced to call them supergiants. These are objects that we now know to be the bulbous evolutionary fate that awaits high-mass stars. Normal main sequence stars such as the Sun are officially called dwarfs, which is clearly what they would look like to a giant; dwarf stars typically have a million times smaller volume than giants. Yet let us not confuse normal dwarfs with the hot degenerate stellar corpses that we call white dwarfs, which have a million times smaller volume than normal dwarfs.

Note the rapid loss of descriptive adjectives at the "dwarf" end. I am convinced that it is the result of the relative scarcity of English words that describe what is smaller than normal when compared to words that describe what is bigger than normal.

The day that *super-* becomes an insufficient modifier, astronomers will be armed and ready. Some of us have reserved the term *super-duper* for the occasion.

Alphabet Soup

Astronomers have always had a penchant for lettering things. Ever since Joseph Fraunhofer lettered major features in the solar spectrum in the early 1800s, astronomers have been lettering things from stellar surface temperature to galaxy shapes. Some of Fraunhofer's nomenclature is still used today to identify strong absorption features: atmospheric "A" and "B" bands, sodium "D," calcium "H and K," and the "G" band of calcium hydride.

As is detailed in chapter 9, the lettering tradition continued across the turn of the century when Annie Jump Cannon at the Harvard College Observatory classified and sequenced stellar spectra according to the strength of an absorption feature due to hydrogen. The stars with the strongest features were lettered "A," the stars with the next strongest features were lettered "B," and so forth. It was later found that a temperature (color) sequence revealed more stellar physics than a spectral line-strength sequence. Some lettered categories were discarded. Others were combined. What remains is the famous spectral classification sequence that is still used today to classify all stars. In order of decreasing temperature we have: O B A F G K M. This sequence has occupied, and will continue to occupy, the minds of mnemonic writers for decades.

Letters are also used to convey shapes. In 1925 the American astronomer Edwin Hubble classified the appearance of galaxies in a lettering scheme that still bears his name. It is this lettering scheme that one follows as you move from "early galaxies" to "late galaxies" along Hubble's tuning fork diagram. Preserving the I-call-them-as-I-see-them tradition of astronomers and baseball umpires, Hubble identified elliptically shaped galaxies with the letter "E"; the most round among them was labeled E0 (pronounced "E-zero"), while the most elongated among them was labeled E7. Hubble labeled flat, spiral-shaped galaxies with an "S." If the spiral arms were connected by a straight barlike section in the middle of the galaxy (as is true for nearly half of all spiral galaxies), then a 'B' was appended to the 'S.' Some spirals were so puffy-looking that they resembled elliptical galaxies. These became their own category called S0 (pronounced "S-zero"). Tightly wound spirals were sublabeled "a," intermediate spirals "b," and loosely wound spirals "c." In modern times this three-tiered scale was expanded to describe *really* loose spirals, which are sublabeled "d." For symmetry with the "SB" of barred spirals, nonbarred spirals are now noted by "SA." And, of course, irregularly shaped galaxies are labeled "Irr." If you are curious, the family photo of the Milky Way galaxy and its nearest neighbors would show: the Milky Way—SAbc (a cross between types b and c); the Large Magellanic Cloud—Irr; the Small Magellanic Cloud—Irr; the Andromeda galaxy—SAb; and NGC205 (a satellite galaxy to Andromeda)—E5. As you can begin to see, Hubble's original scheme is now extended to describe all sorts of galaxy morphology. My favorite among them is the letter "p," which you add to the classification if, no matter how else you describe it, the galaxy just looks peculiar.

Stars that vary in luminosity are no strangers to lettering schemes. Omitting A through Q, the first variable star discovered in a constellation is noted by R followed by the genitive of the constellation name. Clearly, only a few variable stars can be discovered before one exhausts the alphabet. By convention, after Z comes RR, then RS, and so forth, all the way to RZ. If that's not enough, then the scheme resumes at SS, then ST, and so forth all the way to SZ. This continues until ZZ. If the constellation is big and has many stars, then it may need even more letter combinations than those up to ZZ. When this happens, the scheme continues at AA, then AB, through to AZ. Next comes BB, then BC, through to BZ. The last possible lettered variable star is QZ, because afterward you would hit RR, which was already used after Z. This naming scheme, for no particular reason, ensures that the first letter is always earlier in the alphabet than the second letter—unless the letters are the same. One final criterion is that the relatively modern letter "J" is never used. If you kept count of all this, then you should have obtained 334 combinations.

If a constellation has the audacity to exhaust this many letters and letter-pairs, then stars are simply numbered (not from one, but from the number that is appropriate if all previous variables in the constellation were numbered instead of lettered) with a prefix of "V" for variable. For example, the star V471 Tauri is a well-studied variable that can change its brightness abruptly. If a variable star is discovered to be the prototype of a new class of variable stars, then the entire class is named for that star. The famous star RR Lyrae (discovered after, of course, Y Lyrae and Z Lyrae), in the constellation Lyra, defined the properties of what are now called "RR Lyrae" variables.

If you think the lettering scheme for variable stars is obtuse, then you might as well skip over the next paragraph on asteroids.

Asteroids are lettered in order of discovery. Built into the lettering scheme, however, is a time-of-year indicator. If you split the year into twenty-four semimonths and letter them "A" through "Y" (omitting "I," keeping "J," and never reaching "Z"), then you have the first letter of a newly discovered asteroid at that time of year. For example, the first letter of an asteroid discovered at any time during the first half of January is "A." The second half of January would yield asteroids with "B" as a first letter, and so forth down the calendar. For each semimonth, newly discovered asteroids have their second letter sequenced "A" though "Z" (omitting "I," once again). In case you missed it in the news, Earth had a close encounter with a 200-million ton asteroid 1989FC, which was the third asteroid discovered in the second half of March in the year 1989. In a semimonth where more than twenty-five asteroids are discovered, the second-letter sequence is restarted with an appended numeral. For example, the asteroid 1980RZ was followed by 1980RA1, and then by 1980RB1.

Comets are also each lettered in annual sequence, except that the scheme is considerably less ornate, and all twenty-six letters of the modern alphabet are used. You simply letter comets in order of discovery in a year, and double back with an attached numeral when you run out of letters: Using lower case letters, begin with "a" and continue through "z." Resume with "a1" through "z1," then "a2" through "z2," and so forth.

Supernovae are also simply lettered, but the first pass uses uppercase and the second pass uses a double alphabet in lowercase. In other words, begin with "A" through "Z," and continue with "aa" through "az," then "ba" through "bz," and so forth. The famous supernova that was discovered in the Large Magellanic Cloud in 1987 (which made the cover story of Time Magazine with the headline "BANG") was the first supernova to be discovered that year. Its official designation is "1987A." The ultimate "zz" has never been reached, so we do not yet need to worry about what happens afterward.

Unlike supernovae, which keep their lettered identification forever, asteroids graduate to number-name status after their orbits and identities are confirmed. The number part is simply a catalogue sequence, while the name part is assigned by the discoverer in honor of any person, place, or thing. For example, the asteroid 2873 Binzel is named for Rick

Binzel, a good friend friend and colleague of mine who has built his professional career on the study of asteroids. The asteroid 517 Edith is named after somebody I've never met named Edith. The asteroid 2906 Caltech is named for the California Institute of Technology, and 1432 Ethiopia is named for the African nation. But there is also the asteroid 1896 Beer, which is very well known to some.

Comets, like asteroids, get named when their orbits and identities are confirmed, except that they are never named for places or things—or alcoholic beverages. They are named only for their discoverers or for the first person to compute a reliable orbit. The new label also identifies the year and the sequence (in Roman numerals) of a comet's closest approach to the Sun. Famous examples include comets 1965VIII Ikeya-Seki, 1970II Bennett, and 1973XII Kohoutek. To distinguish short-term periodic comets from those comets whose orbits are longer than anybody is willing to document, the letter "P" is often inserted. For example, during its most recent visit to the inner solar system, Halley's comet was officially, and unpoetically, designated 1986III P/Halley.

Roman Numeral Soup

There are Type I, Type II, and Type III comet tails; there are Type I and Type II supernovae; there are Seyfert galaxies of Type I and Type II; there are Population I and Population II stars; and there are stellar luminosity classes of Type I through VII. There is nothing mysterious about these classifications. They are the product of a humble attempt to distinguish more than one variety of object in a given category.

A picture-book that compares all the comet tails of recorded history would reveal a diversity that is no less rich than that found among fingerprints or snowflakes. In science, however, simple appearance is rarely as important as substance. Comet tail taxonomy recognizes only three basic types. The Type I tail is composed primarily of molecules whose electrons have been stripped away to form ions. It is commonly known as a "plasma tail." The Type II variety is less exotically composed of small dust or ice particles and is known simply as a "dusty tail." A comet's tail always points away from the Sun—except when it doesn't. There is usually something anomalous about such a tail, so astronomers call them "anomalous tails" and tag them Type III. Anomalous tails are typically caused by the chance alignment of the comet with illuminated debris in the plane of its orbit. Comets commonly display all three tail types simultaneously, which handily contributes to their uniqueness.

All supernovae have at least one property in common: a star explodes. If you wish to understand supernovae in detail, however, then further classification is warranted. Type I supernovae have weak hydrogen absorption features in their optical spectrum while in Type II supernovae these features are strong. Recently, Type I supernovae, based on closer examination of the class, have been split into two categories, Type Ia and Type Ib. This schism helped to reveal that Type Ib *and* Type II owe their origin to the explosive death of an isolated high-mass star. A Type Ia supernova, however, is the consequence of mass transfer in a binary system where a white dwarf recipient explosively unbinds from a thermonuclear runaway.

Seyfert galaxies are normal-looking spiral galaxies with remarkably luminous nuclei. Carl Seyfert first identified the class in 1943 as part of a larger survey of spiral galaxies. Once again, the class subdivision is based on the appearance of hydrogen in the spectra. Type I Seyfert galaxies have much stronger hydrogen emission than Type II Seyferts.

The light from elliptical galaxies is dominated by old red stars while the light from spiral galaxies is dominated by young blue stars. This simple observation leads to the idea that ellipticals and spirals have different stellar populations. The most recently formed stars are called Population I. They have been enriched (or polluted, if you prefer) by heavy elements that have been scattered through space by previous generations of supernovae. The oldest stars, however, were born before significant enrichment could occur—they are called Population II. Ellipticals are generally considered to be Population II while spirals have a mix of Population II and Population I. The population concept is only a convenience that actually clouds the reality of transitionary populations within spiral galaxies. To confuse matters further, note that Type II supernovae are found only among Population I systems.

Luminosity class is one of the few intuitive Roman numeral classification schemes. In basic terms it is an indicator of how big a star is. Class I are supergiants. These are stars that can get as big as the orbit of Mars. (That's why they are called *super*giants.) Class III are normal red giants, and Class V are main sequence "dwarf" stars, like the Sun. The smallest are among Class VII, which are exclusively white dwarfs. The three other classes are intermediate in size: Class II contains bright giants, Class IV contains subgiants, and Class VI contains subdwarfs.

Greek Soup

The eighty-eight constellations in the sky have their brightest stars lettered in order of brightness. The squiggly looking, lowercase, twenty-four-letter Greek alphabet ($\alpha\beta\gamma\delta\epsilon\zeta\eta\theta\iota\kappa\lambda\mu\nu\xi o\pi\rho\sigma\tau\upsilon\varphi\chi\psi\omega$) has been endowed with this honor. The brightest star in any constellation is named with the first Greek letter α (alpha) followed by the genitive case of the constellation name. Dimmer stars are named in sequence down the alphabet. There exist notable exceptions to this rule, however. Some constellations contain stars of approximately equal brightness, such as the seven brightest stars in Ursa Majoris. These magnificent seven form the Big Dipper and are simply lettered in sequence from west to east across the sky.

Other famous stars include α-Centauri in the southern constellation Centaurus, which happens to be the closest star system to the Sun, and β-Cygni, which is also known as Alberio, a beautiful double star system in the northern constellation Cygnus. The science fiction series *Star Trek* borrowed this nomenclature and appended a Roman numeral to indicate a planet's number according to its distance from a star. One of their better-known planets is α-Ceti-V to where Khan (the bad guy) was banished.

Catalogue Queries

The astronomer's cosmic laboratory contains billions of stars and galaxies. It should be no surprise that catalogues proliferate the profession. There are three basic naming formats. One scheme uses somebody's name followed by a number like Messier 101, or Arp 337. These are objects that have simply been collected together in a list and then numbered. The Messier catalogue happens to be a list of fuzzy objects in the sky that was originally intended to prevent confusion with what might otherwise be a newly discovered comet. The Arp catalogue is a list of peculiar-looking galaxies, most of which are gravitationally disturbed by a near neighbor. Occasionally, an astronomer (or an institution) will publish more than one freshly numbered list. These objects require an extra identifier. For example, II-Zwicky-70 (a compact, irregular galaxy) is the seventieth object on Zwicky's second list of compact objects, and 3C273 (the brightest and first confirmed quasar) is the two hundredth and seventy-third object in the third University of Cambridge catalogue of radio sources.

Another basic scheme uses a name followed by the approximate coordinates of the object on the celestial sphere. For example, IRAS 1243 + 30 is simply an object located at 12 hours 43 minutes in right ascension and + 30 degrees in declination that was discovered by the *Infrared Astronomical Satellite*. Cosmic objects that are not fortunate enough to make it into anybody's list are simply noted with their coordinates preceded by the letter *A* for "anonymous." Among astronomers, this coordinate designation is affectionately referred to as the object's telephone number.

The third basic scheme is a hybrid of the first two. Here all objects are listed in order of increasing right ascension and are numbered in this order. Famous (enormous) catalogues like the New General Catalogue of Non-stellar Objects (NGC) and the Smithsonian Astrophysical Observatory Star Catalogue (SAO) are well-known examples. Incidentally, NGC 224 is the Andromeda galaxy and SAO 000308 is Polaris, the North Star. And if you see a star labeled BS 1457, don't be alarmed; it's just a star (in the constellation Taurus) from the Yale Bright Star Catalogue.

Epilogue

Among its varied and numerous duties, it is the job of the International Astronomical Union (IAU) to establish rules of naming and nomenclature. In many cases, however, these rules are set *after* a naming scheme or term has already been widely used by professional astronomers. Consequently, typical rules of the IAU are simply the formal recognition of naming trends. This approach to the jargon of a discipline tends to preserve the history, spontaneity, and novelty of the language of scientific discourse. These ingredients are likely to ensure that the discoveries of astronomy will forever remain attractive and accessible to the general public. It is also no surprise that astronomy, the oldest of all sciences, is the most frequently tapped discipline for science fiction literature and films. The subject, as well as the terms themselves, seem to capture the imagination and romance of scientific exploration.

May the terms be with you.

PART TWO

Some Unifying Ideas in the Physical Universe

· 5 ·
Center of Mass

Your center of mass is a place you cannot visit but you always carry with you. Like memories, it is part of life's baggage.

☙

*T*he center of mass of a uniform spherical object is found, quite sensibly, at its center. For squishy and flexible objects that are not spherical, like most humans, the center of mass can be inside or outside the object. With some clever contortions you can do some interesting things with your center of mass. For example, the best way to rid your body of its center of mass is to bend forward at the waist. If you bend far enough then your center of mass will emerge from your midsection and be suspended somewhere over your feet. Unless your feet are nailed to the ground, you will fall over the moment your center of mass extends beyond your toes. Notice that because your feet do not extend behind you, it is quite easy to fall backward. If you bend that way, then your center of mass emerges from your lower back with no supporting feet and toes below it.

You may have noticed that high jumpers and pole-vaulters curl around the bar as they pass over it. As a consequence, their center of mass leaves their bodies and actually sails *below* the bar. This little-known fact is the secret of these two track and field events. Olympic history was made at Mexico City in 1968 when the American high jumper Dick Fosbury set a world record by jumping "backward" over the bar with his back toward the ground. This clever innovation allowed Fosbury to send his center of mass farther below the bar than is possible by a simple forward jump. The entire spinal column of any flexible athlete can bend backward almost as easily as it can bend forward. You may have discovered, however, that your knees (whether or not you are an athlete) bend only one way. They bend in the way that will allow a backward jump to separate the center of mass from the body by the farthest amount. If you jump forward over the bar, then you do not get to bend at your knees—unless you have no bones in your legs. One could say that Dick Fosbury jumped higher than ever before simply by jumping lower than ever before. One can apply the same physical reasoning to the pole vault. Pole-vaulters already have their back toward the ground from the moment they are airborne. There is no reason why the world record could not improve immediately if vaulters managed to continue backward over the bar.

The Moon's gravity is about one-sixth Earth's gravity. Naive thinking would suggest that a high jumper would jump six times higher on the Moon. But this is true only from the point of view of the center of mass. The center of mass of a tall high jumper starts at about four feet above ground level and, at the peak of the jump, comes within about a foot of the bar. On Earth, a world record jump clears eight feet. If you add and subtract the right numbers you will find that the high jumper's center of mass was raised a mere three feet. The same jump on the Moon will also start with a center of mass four feet above the ground, and it will also come within about one foot of the bar. It is only the in-between distance that benefits from the Moon's reduced gravity. The three-foot rise of the high jumper's center of mass will, on the Moon, sail six times higher—to eighteen feet. Once again, if you add and subtract the right numbers you find that the world record 8-foot jump becomes a Moon-record 23-foot jump.

Pole-vaulters also start with their center of mass about four feet above ground and also clear the bar with their center of mass sailing about a foot below it. But their center of mass travels much farther. By the same arguments used for high jumpers, we find that Moon pole-vaulters gain considerably. A 19-foot pole vault on Earth becomes an 89-foot vault on the Moon. The nursery rhyme "and the cow jumped over the Moon" might be best reworded "and the cow pole-vaulted over the Moon."

Center of mass is one of the most widely used concepts in physics. It is computed routinely for subatomic particle interactions, and it occupies a pivotal role if one wishes to understand orbits of binary stars and binary galaxies. Sir Isaac Newton, the champion of gravity and apples, provided early insight to the related concept of center of gravity in the *Principia* (Book I, Corollary IV):

> The common centre of gravity of two or more bodies does not alter its state of motion or rest by the actions of the bodies among themselves, and therefore the common cen-tre of gravity of all bodies acting upon each other (exclud-ing external actions and impediments) is either at rest or moves uniformly in a straight line.

If two subatomic particles (or two of anything) are in motion, we can refer to the center of mass of the individual particles or, if we choose, we can refer to the center of mass of both particles combined. If the particles have equal mass and equal but opposite velocities, then the center of mass of both particles combined is exactly midway between them and stationary. If the particles had, say, kinetic-energy-absorbing chewing gum affixed to their front surface, then they would collide, stick together, and stop.

The combined center of mass would not be stationary if one of the particles had more mass. Given the same chewing gum setup, the particles would collide, stick together, and would still be moving (though more slowly) in the same direction as that of the high-mass particle before the encoun-ter. From the low-mass particle's perspective, it had its for-ward motion halted and then reversed.

To have your forward motion abruptly halted and then

reversed can be traumatic if you are bigger than a subatomic particle. In automobile accidents, the safest vehicles on the road are the big, 4-door, 8-cylinder, 10-miles-per-gallon American cars of yesteryear. Actually, the safest vehicles are probably fully loaded cement trucks, but these are not common participants in road accidents. In a head-on collision between our vintage American car and a small imported car, it is very likely that the imported car would have its forward motion abruptly halted and then reversed while the American car's velocity would become only somewhat lower. A safety note: Passengers without seat belts do not get their forward motion slowed, halted, or reversed—they simply sail into, or through, the windshield.

The location of the center of mass between two bodies is easy to find. For example, Earth is eighty-one times more massive than the Moon. The center of mass is therefore $1/81$st of the way from Earth's center to the Moon's center, which would put it about one thousand miles beneath Earth's surface along an imaginary line that connects with the Moon. We, as earthlings, mistakenly claim that the Moon orbits Earth. To be precise, one must say that Earth and the Moon orbit around their common center of mass. Gravity and orbit dynamics dictate that the Moon *and* Earth are always opposite each other with the center of mass between them. This means that the Moon and Earth circumnavigate their joint center of mass once every lunar month. In turn, the Earth-Moon center of mass traces annual orderly ellipses around its mutual center of mass with the Sun.

The Sun is two hundred thousand times more massive than the Earth-Moon system, so the Earth-Moon-Sun center of mass is $1/200{,}000$th of the way from the center of the Sun to the Earth-Moon center of mass. This is just over four hundred miles from the Sun's center—deep within the core of thermonuclear fusion. The planet Jupiter is more massive than all other planets combined and is farther from the Sun than Earth. These factors put the Jupiter-Sun center of mass much farther than four hundred miles from the Sun's center. With nine or ten orbiting planets one can imagine the complexities with tracking the Sun's motion. In spite of this, the enormity of the Sun's mass ensures that it contains, deep within its surface, the single center of mass of the entire solar system.

Sir Isaac first showed that for spherical objects (or for any shaped object if you were far enough away) the force of gravity acts as if all the mass were located at its center of mass. Returning to Newton's *Principia* (Book I), he states:

> If spheres be however dissimilar . . . I say, that the whole force with which one of these spheres attracts the other will be inversely proportional to the square of the distance of the centres.

This discovery is more profound than it might sound. For example, it says that the Sun's force of gravity as felt by Earth is unrelated to the size of the Sun. If the Sun were to swell up like a beach ball or shrink to become a black hole, then life, of course, would be very different (or nonexistent), but planet Earth would still complete its orbit in 365¼ days and keep its average orbital distance of 93 million miles.

Current theories of stellar evolution do not predict a black hole as the ultimate fate of the Sun. We do, however, expect an occasional black hole to be found among the 50 percent of all stars that are in binary systems. If one star out of a happily orbiting pair collapses to a black hole, the pair will remain in happy mutual orbit. If you follow their common center of mass through space, then what used to be two stars dancing loops would become one star dancing what appears to be lonely loops. Indeed, this is the most reliable means by which underluminous companions are discovered.

In addition to black holes, the list includes neutron stars, white dwarfs, "brown" dwarfs, and Jupiter-like planets.

The center of mass of a system, when stationary, cannot set itself into motion. For example, a motionless rocket in space has a motionless center of mass. If the rocket wants to go forward, it cannot bring its center of mass with it. The rocket must aim and jettison a piece of itself, such as combustive fuel, in the opposite direction. For the remainder of the rocket trip, no matter where the rocket goes, the center of mass of the rocket *plus* the jettisoned fuel will not have budged from its original motionless spot in space. A survival note: If you are ever stranded in space without fuel, then a sequence of well-aimed flatulents will send you on your way again.

If you do not like math or fractions or numbers but you want to find the centers of mass of things, then all you need to do is poke around for balance-points. An unevenly loaded dumbbell weight-set will only balance if lifted at its center of mass, which would obviously be found closer to the side of the bar with the greatest weight.

Seesaws (or teeter-totters) may have provided the first occasion where you as a child noticed that some of your friends did not weigh the same as you. If you were a little kid, or simply underfed, then it was no doubt unforgettable the first time you were kept stranded some four feet above the ground with your legs dangling helplessly below you, while your bigger or overfed friend sat firmly planted at the other end. If you were lucky, your playground's seesaw had an adjustable point of balance that allowed small children to see and saw (or teeter and totter) with bigger children—little did you realize that you were finding the center of mass between you and your friend. Even less did you realize that if you were isolated in space, you could orbit each other around that same spot.

The center of mass of tossed objects, no matter the shape, will follow smooth, arched trajectories. The wobble-flopping that some objects are known to display (like jugglers' bowling pins, boomerangs, and airborne baseball bats) is a visual consequence of the object's off-center center of mass. A slow-motion film that follows the center of mass will show the smooth path that we have learned to expect from normal spherical objects. In fact, if the jugglers' bowling pins were not strongly bottom-heavy, then they would be much more difficult to catch. Freely rotating or flipping objects must do so around their center of mass. The bowling pin's design permits the neck of the pin to flip securely into your extended palm without the fat bottom getting in the way.

The only object in the universe that exists entirely *at* its center of mass is the remnant of the death of a high-mass star. There is no known force in the universe that can prevent the collapse of these short-lived luminaries. The collapse is expected to continue until all the star's material occupies its own center of mass. Light cannot escape these objects. If you fall in, you don't come out. This is why we call them, quite sensibly, black holes.

A household cat knows all about its center of mass. If you have ever seen a cat accidentally fall or have been deranged enough to toss a cat through the air, you may have noticed that it usually lands on its paws. This feline feat of acrobatics may also hold for the larger jungle cats, but I have never tossed a tiger to test this hypothesis. Upon being thrown, an airborne cat will use its momentum of rotation to flip its body (paws included) around its own center of mass. Just as rotating ice skaters spin faster upon drawing their arms closer to their body, a tossed cat will curl its body to rotate quickly into position.

The moment the cat's paws point toward the ground, the cat immediately unfolds its body, which abruptly slows its rotation and allows it to land on its paws. Clearly, cats who know one or two laws of physics will earn the full nine lives that are due their species. This impressive maneuver happens quickly, but not quickly enough for you to grab the cat and slam it into the ground. This is the sort of thing that professional wrestlers do to each other, but it is not a fitting treatment for a cat.

The center-of-mass concept can be generalized to answer questions such as, "Where is the center of mass of the continental United States?" If you cut from a sheet of cardboard the exact shape of the contiguous United States, then by poking around underneath the cutout you could find the exact spot (there is only one) where the United States balances. If you are curious (but you don't feel like carving your expensive road atlas), then the spot lies in the state of Kansas about fifteen miles south of the Kirwin Reservoir off the north fork of the Solomon River in Rooks County. If you feel that you must build mountain ranges into the cardboard before you poke, then make sure they have the correct relative height to the size of the cutout. If the United States were cut from an 8½ x 11 inch sheet of cardboard, then the Rocky Mountains would loom about one-hundredth of an inch above the page.

Every ten years the U.S. Census Bureau counts people and determines where they live. One can then ask, "Where is the center of the population of the United States?" One could have equivalently asked, "Where is the population's center of mass?" For simplicity's sake, if we assume everybody has the same mass, then imagine a (very large) cardboard cutout of the United States with all residents standing where they live. The Northeast would be quite dense with its string of large cities: Washington, D.C., Baltimore, Philadelphia, New York City, and Boston—all within 450 miles.

The upper Midwest farming states, such as North and South Dakota, Nebraska, Kansas, and Iowa, of course have a relatively low population density, as does the entire set of states in the Rocky Mountain time zone. The population of California increased sevenfold in the first half of the twentieth century, which helped to yank the nation's center of mass from the crowded grips of the East Coast. The general population shift in the United States forms a major part of professional demographic studies.

An object's center of mass under the influence of gravity will always try to occupy a lower position. The breakfast cereal box translation would be, "Some settling of contents may have occurred." Another interpretation would be, "Top-heavy things tend to topple." Indeed, there is no shortage of proverbs in the world of physics and cereal boxes. Intuition and physics tell us that objects with low centers of mass, like sports cars and turtles, are stable. Conversely, objects with high centers of mass, such as triple-decker buses and coconut palm trees, tend to lean. Part of the purpose of a deep hull in a seagoing ship is to place the ship's center of mass as low as possible. In most seagoing vessels the center of mass is below the water level. Exceptions include catamarans and rafts. Like an inverted pyramid, a perfectly capsized ship is an extremely unstable and unlikely configuration. This fact of physics did not seem to be a concern to the makers of the 1972 disaster film *The Poseidon Adventure*, where a tidal wave capsizes a large ocean liner, and the survivors have to escape "up" through the hull.

In a letter to the Westminster scholar Richard Bently in 1692, Newton reasoned that if the universe always existed then the universe must be infinite because if it were not, then the collective gravity would draw the matter to "fall down into the middle of the whole space and there compose one great spherical mass." No need to worry about a center-of-gravity apocalypse just yet, however. Recent evidence suggests that not only is the universe finite in size but it did not always exist and is likely to expand forever.

As already noted, center of gravity is a concept that is closely related to center of mass. In everyday life, the two terms are interchangeable, but strictly speaking they are never in exactly the same place in your body. On a normal day, your center of gravity is slightly offset in the direction of Earth (or in the direction of the nearest source of strong gravity). Earth's gravity is stronger on your lower half because that half is closer to the center of Earth than is your upper half. The center of gravity is therefore shifted slightly toward Earth. Obviously, the difference in Earth gravity across your height is small. This is why center of mass and center of gravity are often thought to be in the same place. There are locations in the universe, however, where your center of mass and your center of gravity are widely separated. If you fall feet first into a small black hole—do not try this at home—you will notice that the force of gravity changes quickly. Your center of gravity (formerly in your midsection) will systematically separate from your center of mass as it approaches your feet. It is not clear what this will feel like. But as the difference in gravity between your head and your toes rips you apart, your body segments will descend, and be squished, until they occupy the exact spot that is the black hole's center of mass. Your only pleasure will be derived from realizing that rather than just reading about a center of mass—you actually became a center of mass.

· 6 ·

Energy

One of the greatest triumphs of eighteenth- and nineteenth-century physics was the formal understanding of heat energy and its interchangeability with mechanical energy. Out of these efforts was born the branch of physics called "thermodynamics," which was pioneered through the efforts of many scientists, including the Scottish engineer James Watt (born 1736), who perfected the modern condensing steam engine; the American physicist Benjamin Thompson (born 1753, later Count Rumford), who first proposed that heat is a form of energy; the French engineer Nicolas Léonard Sadi Carnot (born 1796), who provided the first analysis of heat engines; the British physicist James Joule (born 1818), who performed careful experiments to prove that heat is, indeed, a form of energy; and the British physicist William Thompson (born 1824, later Lord Kelvin), who helped to formulate a consistent physical theory.

Modern society owes its industrial success primarily to the invented machines that allow work to be accomplished from energy that is *not* supplied by the physical labor of humans or of other animals. It is no accident that the nineteenth-century industrial revolution coincided with the development of thermodynamics. A curious twentieth-century analog is that computers allow certain computational tasks to be completed without the intellectual labor of humans, so that society can now substitute machines for both our bodies and our brains.

Meanwhile, back in the rest of the cosmos, the conversion of one form of energy to another plays a major role in stellar evolution, stellar orbits, and in the fate of the universe itself.

℃℃

*T*here are many different types of energy, although not all of them manifest themselves in everyday life. Among those that do, there is one type of energy that kills more people per year than any other. It is the energy you have by simply being in motion, which is known in the world of physics as your "kinetic" energy.

When you start your car and accelerate to 50 mph onto the freeway and drive for a few hundred miles, you may notice that there is less fuel in your tank than when you started. During your trip, you converted the stored chemical energy of the gasoline into heat energy from the friction of the car's internal moving parts, and into the kinetic of your entire car plus its occupants. When you apply your brakes to return to zero miles per hour, your car's kinetic energy must go somewhere. It transforms to heat by way of the friction between your brake pads and your wheels, and if you skid, between your rubber tires and the road.

In a head-on collision, you also slow, for example, from 50 mph to zero miles per hour, except that this does not happen with the help of your brakes. The kinetic energy of car-plus-driver at 50 mph must go somewhere. It becomes the sole source of energy for the deafening sound of the collision, the crunch of the car's front end, the smashed face and skull of any unseatbelted passenger, the damaged guard rail alongside the road, and any toppled lamp posts. The kinetic energy wielded by an object depends on its mass and on its velocity. But it only takes a small change in velocity to induce a big change in the kinetic energy. More precisely stated, the kinetic energy depends on the mass and on the square of the velocity

$$Kinetic\ Energy\ =\ {}^{1}\!/_{2}\ \times\ mass\ \times\ velocity^2$$

This formula, translated into a proverb, would read, "Speed kills." A sobering example is that at 70 mph you have nearly *twice* the kinetic energy of what you had at 50 mph. In other words, if you were to drive 70 mph rather than 50 mph, then every aspect of a car accident would be twice as destructive. Not only would the sounds be louder, but, on average, the damage to your car would be twice as extensive, and you would be twice as likely to die. Yes, speed does kill.

While departments of transportation try to help people stay alive on the highways, the U.S. Department of Defense tries to find ways to kill people. Using the principle that speed kills, a rifle was invented that fires a relatively small bullet (0.22 caliber), but achieves a muzzle velocity of 3,250 feet per second, which is about three times the speed of sound—you would be hit with the bullet before you could hear the rifle shot. This weapon is the M-16 assault rifle designed by Eugene Stoner in 1959, which was widely used by the American forces in the Vietnam War. It replaced the Thompson submachine gun, which was used throughout the Korean War and which fired relatively slow-moving "fat" (0.45 caliber) bullets. Stoner realized that high muzzle velocity is more important than massive bullets. This physical principle also had not escaped the Russian weapons designer Mikhail Kalashnikov. His AK-47 rifle, the Russian high-velocity counterpart to the M-16, was widely used by the North Vietnamese. It is the kinetic energy of the bullet, obtained from the stored energy of explosive chemicals, that transfers to the target, which in the case of human flesh can be quite devastating. A letter home from Vietnam, written by Army Corporal George Olsen in 1969, contains the following passage:

> We crawled within six feet of one group [of the North Vietnamese Army] and then charged, and all hell broke

loose. . . . One [of them] went down fighting; [he] shot our point man in the ankle at fistfighting range, [but] then was blown apart by the sergeant leading us. I won't go into detail, but it is unbelievable what an M-16 will do to a man—particularly at close range. The only conceivable comparison is swatting a bug with a chain-mail glove. Enough said—perhaps too much.

[Our point man's] wound, of itself, wasn't serious, but the power and shock of a modern rifle bullet is absolutely unbelievable and within two minutes of being hit he was fighting for his life in shock.

Corporal Olsen was killed in action on March 3, 1970.[1]

Chemical energy is not the only way to set something into motion. Gravity is well known for this ability. For example, if you should precariously set a pan of freshly baked peach cobbler to cool on the narrow sill of the open window of your eighth-floor apartment, and if, by chance, the peach cobbler should fall out, then it will increase speed (gain kinetic energy) all the way to the ground. Unless you have lived in the basement all your life, this airborne fate of your peach cobbler should come as no surprise. What was the source of energy that became the kinetic energy of the cobbler? It was not gasoline. We presume it was not gunpowder. Rather, it was you (or perhaps your elevator). You carried the peaches. You carried the flour. You carried the brown sugar. You carried the butter. You carried all these ingredients from ground level to the eighth floor *against the will of gravity*. This common consequence of a shopping trip endows your food with the potential to recover the work you did against gravity. In genuine terms of physics, the food was given gravitational potential energy simply by be-

1. In Bernard Edelman, ed., *Dear America: Letters Home from Vietnam* (New York: Norton, 1985), pp. 64–65.

ing lifted to some height above the ground. The higher above the ground the food is taken, the more potential energy it gains, and the faster it will hit the ground after it flies out the window. What then happens to the kinetic energy? It promptly explodes the food in a manner that is commonly described with the word "splat." But beware—it also may damage the ground or any unfortunate person below.

On an average day, Earth plows through about 1,000 tons of interplanetary meteors. As they fall toward Earth's surface, most of these meteors lose all their kinetic energy in a spectacular way as shock waves and friction with the atmosphere makes them burn. They become what we all identify as "falling" stars in the nighttime sky. Some of the larger meteors actually survive the trip through Earth's atmosphere and hit the ground with tremendous kinetic energy. What then happens to the kinetic energy? It blasts holes into the ground. The 25,000-year-old Barringer crater near Coon Butte, in Coconino County, Arizona, is an impressive example of a hole "dug" by the impact energy of an iron meteor. It is fourteen football fields in diameter and about five hundred feet deep.

When astronauts reenter Earth's lower atmosphere from orbit, their heat shields get hot. What is not widely appreciated is that these shields are the thermal repositories for the loss of the spacecraft's kinetic energy. Heat shields do not simply serve as protection, they are a way of slowing down. One might even call them "airbrakes."

Spongy objects such as foam, springs, and car airbags make excellent kinetic energy absorbers. If a pole-vaulter landed on a slab of concrete after a 20-foot vault, then the kinetic energy of the fall would fracture bones and rupture body tissue upon impact. This is the "splat effect" that the peach cobbler experienced. Organizers of track and field events wisely place soft fluffy things near the pole vault and high jump to absorb the kinetic energy of impact. The task of absorbing the kinetic energy is then passed from the human body to these spongy oversized pillows, which is why pillows are normally preferred to concrete. What then becomes of this absorbed energy? It is converted immediately to heat within the absorbers and then dissipates to the atmosphere. Springs, however, take longer to convert kinetic energy into heat. If our pole vaulter landed on springs, then the kinetic energy would swap back and forth with the mechanical potential energy of the springs. You would see the pole-vaulter bob up and down until the energy was converted to heat within the springs.

Children's toys are no exception to these rules. When you shove a jack-in-the-box clown back into its box, you provide energy that gets stored in the inner spring. When the lid is released and the clown pops out, the spring converts its stored mechanical energy into the kinetic energy of the clown. Only when the bobbing stops has the spring converted all available energy into heat, which dissipates to the atmosphere. Many toys require you to "wind up" some sort of device that stores mechanical potential energy. The stored energy is then converted to kinetic energy, and the result is a truck that rolls, a robot that walks, or perhaps even a baby doll that pees. In other toys, this mechanical potential energy is converted to sound energy as the robot or doll speaks to you. The only difference between toys that use batteries and toys that need to be wound is that batteries use stored chemical energy obtained from the battery manufacturer, and wind-up toys use stored mechanical energy obtained from you.

The conversion of gravitational potential energy to kinetic energy is a fundamental ingredient in star formation and stellar evolution. In the final collapse of a gas cloud to form a star, there is a precipitous rise in the kinetic energy of the individual atoms of the cloud. Because the cloud is gaseous, the individual atoms cannot fall straight to the cloud's center. Instead, the increase in kinetic energy is revealed through an increase in atomic collisions and an associated increase in temperature. Some of this kinetic energy is also converted to photons of light, which escape into space. Eventually, if the gas cloud contains enough mass, the core temperature will become high enough to trigger thermonuclear fusion.

A similar mechanism allows us to discover the presence of compact cosmic objects with high mass such as neutron stars and black holes. Unlike the Sun, most stars in our galaxy do not travel though space alone. It is not uncommon to find binary, triple, or even quadruple star systems with all members in mutual orbit. If one star first collapses to become a black hole, and another star passes through the red giant phase, then the red giant may fill its Roche lobe and dump matter across its Lagrangian point onto the black hole. Rather than fall straight in, the gaseous matter is likely to spiral toward the black hole's event horizon in a manner not unlike water that runs down a toilet bowl. Friction between the inner, fast-spinning regions and the outer slower-spinning regions heats the gas to enormous temperatures. As a consequence, the funneling gas emits copious quantities of ultraviolet light and X-rays, which is the calling card of a massive yet compact object. Such high-energy emission would be uncharacteristic of an ordinary star. Once again, the gravitational potential energy is converted to the kinetic energy of atomic collisions rather than to the kinetic energy of descent.

There are many astrophysical systems where mechanical energy is not rapidly lost to heat. In clusters of galaxies, for example where there are no galactic airbags or fluffy pillows, galaxies orbit the cluster center with a relatively constant average kinetic energy. For large clusters of hundreds or thousands of galaxies, this average kinetic energy is a direct and reliable measure of the total gravity, which remains the primary means by which the total mass of a galaxy cluster is determined. The same principles of energy and gravity are also invoked to compute the total mass of the larger open star clusters and of all globular star clusters. This method, however, derives total masses for galaxy clusters that are systematically higher (in some cases, by a factor of one hundred) than what you get if you summed the mass of each individual galaxy. The discrepancy was discovered in 1936 by the California Institute of Technology astrophysicist Fritz Zwicky, and it festers to this day as the infamous "missing mass" problem in the universe.

In the reverse of a collapsing gas cloud that gets hotter, the entire universe cools for every moment that it expands. The overall density of energy drops continuously. The temperature of the radiation that permeates all of space, which is the frigid remnant of a hot big bang, is now just under 3 kelvins. If the universe expands forever, then its contents will ultimately meet a cold and dark death as all stars burn out and as the background temperature nears absolute zero.

How much energy does it take to throw a tomato straight up so that it never returns? It may surprise you to learn that the adage "What goes up must come down" is more a statement of human weakness than of the laws of physics. There is, in fact, a particular velocity that an object must have for it to leave Earth and never return. It is called, quite sensibly, the escape velocity. In the absence of atmospheric resistance, Earth's escape velocity is about seven miles per second from the surface, which is 250 times faster than the fastest pitches thrown in professional baseball.

With rockets, or other launch apparatus, however, if you propel a tomato with at least Earth's escape velocity, then you have endowed it with sufficient kinetic energy to leave the force of Earth's gravity forever. Earth's gravity does manage to slow the tomato down somewhat, but you have given it *more* kinetic energy than it would gain had it fallen to Earth from the edge of the universe. In the genuine descriptive terms of physics, the escaping tomato has sufficient energy to climb out of Earth's gravitational potential "well." On this subject, an acquaintance once penned

> Some of what goes up,
> If launched with great ferocity,
>   Will never return—
> It reached escape velocity.
>
> Some of what goes up,
> If propelled both high and far,
>   Burns upon return
> To become a "falling star."
>
> The rest of what goes up,
> Tossed slowly from the ground,
>   Started the old saying,
> "What goes up, must come down!"
>   Merlin of Omniscia[2]

Comets that move with speeds near the local escape velocity of the solar system are only loosely bound to the Sun and may be considered onetime events. Such comets are not uncommon and are often more spectacular than famous ones that are tightly bound such as Halley's Comet. Earth is treated to one or two of the one-timers per decade.

There are four types of orbits that an object can have in a simple gravitational field. If all four varieties are given the same closest approach to the central object, then—se-

2. From *Merlin's Tour of the Universe* (New York: Columbia University Press, 1989), p. 230.

quenced by increasing total energy (potential plus kinetic)—they are: the circle, the ellipse, the parabola, and the hyperbola. If an object's speed is less than the escape velocity, then its orbit will be bound and assume the shape of a circle or of an ellipse. If an object's speed equals the escape velocity, then it will be unbound with a parabolic trajectory. If an object's speed exceeds the escape velocity, then its trajectory will be hyperbolic. The colloquial cool-down phrase, "Don't get hyper!" does have genuine astrophysical relevance. And if "Don't get hyper!" is too strong for your needs, then you can always substitute, "Don't get parabolic!" or "Don't get elliptical!"

For elliptical orbits, or more generally, for any orbit where the orbit distance varies, there is a continual exchange between an object's kinetic energy and its gravitational potential energy. As the orbiting object moves closer, gravitational potential energy gets converted to kinetic energy—the object moves faster. This is precisely what happened with our defenestrated peach cobbler. In orbit, however, the object gets to move farther away again as some of its kinetic energy is converted back to gravitational potential energy. Amusement park roller coasters are actual physics experiments on the conversion of gravitational potential energy to kinetic energy. In a typical gravity-driven roller coaster, the connected cabs are first dragged up to the highest point in the entire ride, which supplies the requisite gravitational potential energy to avoid getting stuck somewhere between two hills. Now comes the physics experiment: the cabs roll down-and-up and down-and-up and down-and-up in a continual exchange of potential energy with kinetic energy. If there were no friction between the cabs and air and between the cabs and the track, then the roller coaster ride would continue forever. But the roller coaster owner depends on this friction to convert your kinetic energy into heat. The

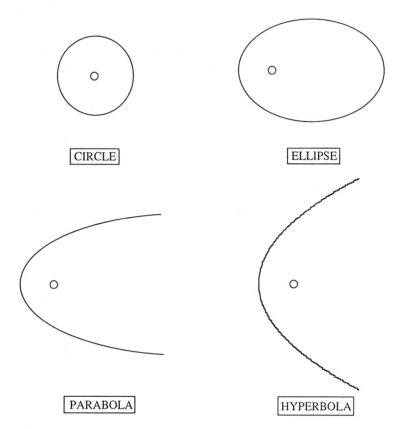

Figure 6.1. Four simple orbits. An object with a circular or elliptical orbit is "bound" to the system. Ellipses can vary in eccentricity from extremely flattened to circular. All planets orbit in ellipses. An object with a parabolic or hyperbolic orbit is "unbound" from the system and is on a onetime trip. For a parabolic orbit, an object's speed equals the escape velocity, while for an object in hyperbolic orbit, its speed exceeds the escape velocity.

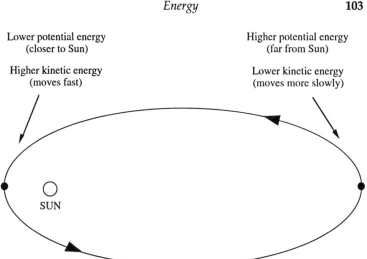

Lower potential energy
(closer to Sun)

Higher kinetic energy
(moves fast)

Higher potential energy
(far from Sun)

Lower kinetic energy
(moves more slowly)

SUN

Figure 6.2. An eccentric orbit around the Sun. There is a continous exchange between the object's gravitational potential energy and its kinetic energy throughout the orbit. When the object is closest to the Sun, it has lost gravitational potential energy in favor of a gain in kinetic energy. Its speed thus increases. The opposite is true for the object at its maximum distance.

successive hills *must* therefore get shorter and shorter, until a short final hill just before the ride ends. If you are a roller coaster enthusiast, then all other things being equal, the roller coaster with the single highest hill will also be the longest and the fastest in the world.

Sunlight is, perhaps, the most pervasive form of energy on Earth. Nearly every form of energy that one encounters on Earth can be traced back to the Sun. A car that runs on rooftop solar panels is, in principle, no different from a car that runs on potatoes. Both use energy derived from the Sun. Wood for your fireplace (or wood in general) can burn because it contains a lifetime of energy that a tree absorbed from the Sun. From the point of view of energy, sitting before a toasty fireplace is no different from sitting before a hearth of sunlight, except that burning logs pollute the atmosphere. Hydroelectric plants derive their energy from falling water, usually through ducts in a dam. They exploit the extra gravitational potential energy that water in the dammed lake has over water in the valley below. But how did the water get from sea level up to the lake in the first place? It is the Sun's energy that helps to evaporate ocean waters, while convection in the atmosphere, which is also caused by the Sun, brings this moisture inland, where it falls out of the sky as rain—indeed, hydroelectric energy is really a form of solar energy.

We can also attribute the complexity of life itself to solar energy. There are countless organic and inorganic chemical reactions on Earth that thrive in the presence of the Sun's abundant energy. How else do you think an acorn becomes an oak tree? If the Sun were to disappear tomorrow, then all flora and fauna would eventually "wind down" until the chemical reactions that sustain life ceased. In addition, all motion would stop as mechanical energy irreversibly converts to heat energy. With the Sun as a rather impressive source of external energy, however, almost anything is possible. And the self-organization of complex forms of matter is *expected*.

For similar reasons, there can never be an isolated "perpetual motion machine," unless you feed it energy—in which case it would be simply be a battery-operated "temporary motion machine." This is not a statement of inadequate engineering; it is a fundamental axiom of the physics of systems that do not tap an external source of energy.

Calories are a direct measure of heat energy. This simple fact seemed to elude the makers of a well-known peanut-filled candy bar in the mid-1980s. The print on the wrapper featured the following absurd claim: "High in energy. Low in calories!" An equally absurd statement might be, "High in weight. Low in pounds!" The human body uses calories derived from food as a means to maintain body temperature and as a source of mechanical energy to do things such as walk, talk, run, circulate blood, and climb stairs. For example, if you just ate a T-bone steak, then the calories you consumed came from the loin of somebody's cud-chewing cow, and the cow was assembled from farm feed such as grass and grain, which was grown with the Sun as a source of energy. Credit the Sun, once again.

An underappreciated aspect of eating cold food is that its net calorie content is always less that what is advertised on the label. Do you want to lose a fast forty calories? Just drink a liter of ice water. Water is often advertised to have zero calories, but by the time it emerges from your body it will have been heated to your body temperature at the expense of your own stored energy. The cost? About forty calories. You get to subtract even more calories for treats such as frozen desserts. Ice cream, for example, is commonly consumed at temperatures well below freezing. Its calorie correction would be quite large. The only disadvantage is that, unlike a liter of water, a tub of premium ice cream packs two or three thousand calories. For this reason, we should not expect the "ice cream diet" to emerge as the latest fad.

Insight to the correspondence between mechanical energy and heat energy was obtained experimentally by the nineteenth-century British physicist James Joule. He revealed that only a small change in temperature results from the dissipation of an enormous amount of mechanical energy. A similar correspondence exists between the food calories that the body consumes and the mechanical energy that is derived from them. In a now famous experiment, Joule stirred a jar of water by the action of falling weights. The gravitational energy of the weights was transferred into the water. Joule describes the experiment:

The paddle moved with great resistance in the can of water, so that the weights (each of four pounds) descended at the slow rate of about one foot per second. The height of the pulleys from the ground was twelve yards, and consequently, when the weights had descended through that distance, they had to be wound up again in order to renew the motion of the paddle. After this operation had been repeated sixteen times, the increase of the temperature of the water was ascertained by means of a very sensible and accurate thermometer. . . .

I may therefore conclude that the existence of an equivalent relation between heat and the ordinary forms of mechanical power is proved. . . . If my views are correct, the temperature of the river Niagara will be raised about one fifth of a degree by its fall of 160 feet.[3]

In a possibly more relevant example than the Niagara Falls, the calorie content from the stored chemical energy in a single McIntosh apple is more than enough for a 150-pound person to climb, against gravity, every step from the ground to the top floor of the tallest building in the world. In overfed nations such as the United States, "calorie" is often taken to be a bad word. However you choose to view it, "calorie" still means energy, even if your body stores excess quantities of it as layers of fat on your tummy.

The calorie content of an apple is not nearly as impressive as the heat content of the world's oceans. The ocean may feel cold when you swim in it, but if you were to add up the vibration energy of *every* water molecule, then you would get an enormous total quantity of heat. In a household example, a standard five-gallon fish tank at room temperature contains over sixty times the total heat energy that is found

3. James Joule, in a letter to the editor, *Philosophical Magazine* 27 (1845): 205; reprinted in Morris H. Shamos, ed., *Great Experiments in Physics: Firsthand Accounts from Galileo to Einstein* (New York: Dover, 1959), p. 170.

in an eight-ounce cup of hot tea. Yes, the cup of tea is hotter, but it contains many fewer water molecules. The tremendous capacity for oceans to store heat energy and influence the local climate is what keeps England from becoming a major cross-country ski resort, even though the entire nation is farther north than the northern tip of the state of Maine. The warm North Atlantic Drift current encircles the British Isles, warms the air, and ensures a relatively temperate climate throughout the year.

Photons of all varieties are also a form of energy. The energy created in the core of the Sun emerges as photons from the solar surface. These photons, however, do not come from chemical energy, or gravitational energy. They are the by-products of thermonuclear fusion, which converts raw matter into energy. Four hydrogen atoms assemble under high pressure and temperature to become a single helium atom. The mass of the helium atom is slightly less than the combined mass of the four hydrogen atoms. The lost mass transforms to energy as described by Albert Einstein's famous formula

$$Energy = mass \times (speed\ of\ light)^2$$

which may be more recognizable when written with its familiar symbols

$$E = mc^2$$

where c stands for the speed of light, which we learned from chapter 3 to be a very large number. A small amount of lost mass, after being multiplied by the square of the speed of light, becomes an enormous amount of energy. For example, just one ounce of matter, converted to energy, could power a 100-watt light bulb for over 800,000 years. This simple and profound fact is why tiny humble atoms can serve as the energy source for nuclear power plants, nuclear bombs, and for every living star in the universe.

There are three ways that heat energy can move from one place to another. One is through "conduction," which is what happens when you hold the fireplace poker too long with the tip embedded among the burning embers. Heat from the fireplace induces the poker's atoms to vibrate faster. These vibrations are communicated systematically up the poker from atom to atom until the top of the poker burns your unsuspecting hand. Conduction is the primary way that solid objects transfer heat.

Another method of heat transfer is "radiation," which simply means energy is transferred directly by photons. Quite independent of your burning hand on the fireplace poker, infrared photons that are emitted from the fire will strike you directly. The human body senses this infrared energy as heat, which is why your exposed skin feels warm when you turn toward a raging fire, yet your skin immediately feels cooler when somebody blocks your view. Photons also travel by radiation from the Sun to Earth along a 500-second journey through interplanetary space. If you had a melt-proof fireplace poker that was 93 million miles long, then you could poke the Sun and tap solar energy by conduction if you felt so inclined. But it is much simpler to wait the few moments for the Sun's photons to arrive.

A third method of heat transfer is "convection." This is how a gaseous or liquid fluid manages to move heat when conduction is ineffective. Returning to our fireplace, we notice that the air nearest the burning embers is at a much higher temperature than the air anywhere else in the room. Much of this hot air convects up the chimney as it is replaced with cooler air along the floor of the room. Unfortunately, the frigid outside air then seeps into your home to replace the hot air that went up the chimney. There is no doubt that a fireplace is a cozy addition to any domicile because of the direct infrared photons it provides. Convection, however, ensures that it does a poor job of raising a room's air temperature.

A pot of water on the stove that is being heated to boil normally sits atop a very high flame or a very hot electrical coil. Rather than communicate this high heat through slow conduction from the bottom of the water to the surface, blobs of steam and pockets of water physically move from the bottom to the top. If this were all that happened, then the water would jump out of the pot and float to the ceiling, which would be in conflict with culinary experience. In fact, blobs of water at the surface descend to the bottom to replace the volume that was previously occupied by the rising blobs. When water behaves this way, it is common to say that the water is "boiling." Raisins make excellent tracers of convecting blobs. Just toss one or two into a pot of boiling water, and you can entertain yourself for hours as you watch them circulate up and down. If you could toss a flame-proof raisin into the Sun, you would discover that convection is the major means by which energy traverses the outer gaseous layers before it is released as photons from the surface.

A few thoughts about these precious solar photons might possibly help you through the workday without caffeine. The next time your energy level is low, or the next time the elevator is broken and you must walk up the steps to your destination, remember that you possess stored chemical energy from the food you have eaten, and that the energy content of the food owes its origin to sunlight. You thus have permission to declare to yourself that you are (indirectly) powered by thermonuclear fusion.

· 7 ·

The Electromagnetic Spectrum

Astronomers extract more information from light than anybody else. The analysis of light—visible and invisible—is the nerve center of the discipline. We collect it, bend it, scatter it, disperse it, reflect it, focus it, split it, image it, polarize it, filter it, and, of course, contemplate it. Sometimes we even tell others about it.

☘

The Astronomer's Family of Photons

It is not widely known that "light" is a broad term that includes the entire family of what is called electromagnetic radiation: radio, microwave, infrared, visible, ultraviolet, X-ray, and gamma ray. One of the triumphs of twentieth-century astronomy has been to build specialized detectors appropriate for each variety of light.

As noted in chapter 3, light moves through the vacuum of space at *exactly* 299,792,458 meters per second, which is just under 186,282.4 miles per second. Light is composed of massless particles called photons. Radio photons, microwave photons, infrared photons, and so forth through to gamma photons differ only in their energy content. Features such as their wavelength and frequency also differ. But by a simple formula, they can each be determined uniquely from the photon's energy.

All photons behave as both waves and particles. When we think of them as a wave, we can then ask about their wavelength and frequency. These are two properties that exist for any type of traveling wave such as sound waves and water waves. It even includes the stand-up-and-sit-down waves that crowds at baseball stadiums are known to produce. The wavelength is, quite reasonably, the distance between successive wave crests (or troughs). The frequency can be measured by counting how many wave crests move by in a chosen time interval such as seconds, minutes, or hours. A common unit of frequency is crests per second, which has been named the "hertz" after Heinrich Hertz, the nineteenth-century physicist who helped to unify the study of the different parts of the spectrum.

Let's visit the extended photon family and see the role they play in our pursuit of what is unknown in the cosmos.

Radio Waves

Radio wave photons have the lowest energy and the longest wavelength of the bunch. Typically, a photon with a wavelength greater than about a foot is a radio photon. Some obvious examples of radio photons include short-wave, television, AM, and FM. You detect these different energy photons by changing channels or tuning to a different station on your receiver. In household radios, the scale on the dial displays frequency in hertz—but it could just as effectively read in units of energy.

Radio photons are emitted by a variety of cosmic phenomena that include radio galaxies, regions of intense star formation, and, perhaps most importantly, ordinary hydrogen atoms.

Radio galaxies are often elliptical galaxies that exhibit evidence for violent activity or explosions at their centers. These explosive regions tend to have strong magnetic fields and a generous supply of free electrons that have become unbound from their host atoms. Electrons (as well as any other charged particle) will spiral and emit light when moving through a magnetic field. The field strengths and the electron speeds that are typical of a radio galaxy ensure that the spiraling electrons emit radio waves known as "synchrotron radiation."

In gas-rich regions of intense star formation, electrons are also readily kicked loose from their host atoms. This time, however, they are recaptured and cascade through a series of energy levels back toward the nucleus. This cascade emits radio photons that readily escape the region.

A lone hydrogen atom (one proton and one electron) in

an interstellar gas cloud will also emit radio waves. This phenomenon was first predicted by the Dutch astronomer H. C. Van de Hulst in 1944 and discovered in 1953 by the American physicists H. I. Ewen and E. M. Purcell. About once every ten million years the spin of the bound electron will flip so that it can be measured to spin in the opposite direction. Each time this happens, the atom loses energy and emits a radio photon with a wavelength of 8.3 inches. In science, of course, we use the metric system, so this is more commonly known as the "21 centimeter radiation." You may wonder if astronomers get bored waiting ten million years to detect this photon. But remember that hydrogen atoms make up more than 90 percent of the atoms in interstellar gas clouds and in the universe itself. There are so many atoms that at any given moment enough electron spins are flipping to keep radio astronomers busy.

The wavelengths of all these radio photons are long enough that they penetrate nearly all intervening material that might otherwise obscure our view. For this reason they can be used to map the star-forming regions and overall structure of gas-rich galaxies, and they permit us to study the structure of the galactic violence at the center and on the outskirts of radio galaxies. In 1960 an object was discovered that was tiny enough on a photograph to resemble an ordinary star, yet it uncharacteristically emitted copious radio waves. It was dubbed a "quasi-stellar radio source," or "quasar" for short. We now know quasars to be the most distant objects in the observable universe. And while most quasars catalogued today are not radio emitters, the name *quasar* is here to stay.

Except perhaps for bad TV sitcoms, which have been thought to drain the brain of all existing intelligence, the photons we call radio waves are the most harmless form of electromagnetic radiation.

Microwaves

Microwave photons are historically part of the radio wave section of the spectrum. As the "micro" in the name implies, they were the smallest of the radio waves. Their wavelengths range from a millimeter or two to about ten centimeters. The microwave window to the universe forms one of our best views of interstellar molecules: carbon dioxide, formaldehyde, and water are just part of a long list of molecules that can absorb and emit microwaves. Because molecules form most readily where atoms are densely packed and slow-moving, we find microwaves to be the best tracer of cold gas clouds.

Atmospheric water, such as clouds or rain, wreaks havoc upon microwave observations. Unlike radio telescopes, microwave telescopes are best located on mountain tops above the lowest cloud layers.

The microwave part of the spectrum also lays claim to the peak energy output from the 15 billion-year-old remnant radiation for the big bang. As noted in chapter 6, the ubiquitous trillion-degree fireball that started the universe has now dropped to a cool 3 kelvins. In 1965 this fireball remnant was first measured in a Nobel Prize-winning observation conducted with a microwave telescope at Bell Telephone Laboratories by Arno Penzias and Robert Wilson. This is the famous "three degree microwave background radiation," although the name is somewhat misleading. As confirmed by one of several famous experiments on the *Cosmic Background Explorer* satellite in the early 1990s, the background temperature is actually closer to 2.7 kelvins. And while the energy remnant from the big bang peaks in the microwave range, it radiates at all wavelengths.

Microwaves are harmless—or perhaps it would be more accurate to say that there is no agreement in the world scientific community about whether low-level microwaves adversely affect humans at all. Microwaves are all around us. They are used for telephone-signal transmitters, satellite relays, airplane radar, walkie-talkies, and many other things, including police "radar guns" in speed traps. Note that contrary to popular parlance, microwaves do not "nuke" food in a microwave oven. As already noted, water has a special relationship with microwave photons. In the presence of microwaves, the water molecule (a common food additive) literally flips. If you flip enough water molecules rapidly—as the high-intensity microwaves in microwave ovens are designed to do—then the friction among them heats the food. Plates and bowls are not made of water. They will, of course, remain cool unless they conduct heat directly from the food.

Infrared

Infrared photons have higher energy and smaller wavelength than microwave photons. Although they cannot be detected by the human retina, we have all had some experience with infrared photons. They are what keeps the French fried potatoes warm for hours before you buy them at your favorite fast-food restaurant chain. They are what you feel when you bring your hand near a hot iron. They are what changes the TV channel when you sit lazily on the couch with a "remote" in your hand.

The wavelength of an infrared photon is long enough (as is true with microwaves and radio waves) to penetrate obscuring clouds in the interstellar medium. Infrared is an excellent probe of heated gaseous regions, especially those that surround beds of star formation. An object's output of energy will peak in the infrared if its temperature is between a few hundred kelvins and about a few thousand kelvins (approximately 50° to 5,000° Fahrenheit). In other words, objects in that temperature range emit more infrared photons than any other variety. Notice that the range includes the human body, heated food, a warm and hot household iron, a fireplace poker (before and after it did some fireplace poking), and a pottery kiln.

In addition to star-formation regions, we expect to find infrared emission from the warm, freshly collapsed disks of material that precedes the formation of planets around a star. This is a difficult measurement to make because usually the host star has already formed and its light (visible and infrared) floods the measurement of the surrounding region. But it has been done. Several such preplanetary disks have been found. Perhaps the most notable among them is Alpha Lyrae, which is also known as Vega, the brightest star in the constellation Lyra.

The surface of the beautiful planet Venus can lay claim as one of the most inhospitable places in the solar system. Its dense atmosphere is over 96 percent carbon dioxide, which is the cause of a runaway greenhouse effect. Light that reaches the planet's surface is absorbed by the rocks and reradiated mostly as infrared. Carbon dioxide *traps* infrared photons. Consequently, the surface temperature has risen to over 900° Fahrenheit—hot enough to melt tin, lead, zinc, and humans. The continued burning of fossil fuels on Earth will double the carbon dioxide content of our atmosphere by the early twenty-first century.

Visible

Human retinal cells of the eye are sensitive to photons with wavelengths from about 4 to 7 ten-thousandths of a millimeter. This range is commonly called visible light. We distinguish these wavelengths by what we call "colors" that are the familiar ones from the rainbow: red, orange, yellow, green, blue, indigo, and violet. These seven siblings are sometimes affectionately and collectively known as ROY G. BIV. Every other color you have ever seen is some combination of these visible light photons.

Molecules and dust in Earth's atmosphere are quite selective about photons that come their way. The particles scatter the shorter-wavelength BIV photons into random directions and ignore the longer-wavelength ROY photons. This fortunate fact of physics operates at all times but shows itself best with the Sun on the horizon. It is here that sunlight passes through the most atmosphere. During sunset or sunrise, ROY comes straight through while BIV is scattered all over the sky. What we get is a photogenic red-orange sun draped within a deep blue-violet sky.

Unlike several soon-to-be-discussed parts of the electromagnetic spectrum, Earth's atmosphere is transparent to visible light. The most widely used tool of astronomers is and has always been the visible light telescopes. Stars such as the Sun have their peak output of energy conveniently placed in the visible part of the spectrum. Actually, any white star such as the Sun emits roughly equal intensity of each color in ROY G. BIV. It then requires a prism or well-placed rain drops to separate the different color photons for the eye to see. As will be detailed in chapter 8, even blue stars (which peak in the ultraviolet) and red stars (which peak in the infrared) emit substantial quantities of visible photons, so that visible light telescopes are overall the best detectors of starlight.

In spite of all this, it is important to realize that visible photons compose a tiny range of all possible photons in the electromagnetic spectrum. As useful as our vision is to life on Earth, we are comparatively blind.

The hydrogen atom, with its lone electron, can produce

features in the visible spectrum of a star that indicate where the electron has been and where it went. Reworded in almost-astrophysical terms, the electron of the hydrogen atom can jump between energy levels that surround the nucleus. Along the way, photons are absorbed (the atom gains energy) or photons are emitted (the atom loses energy). If the surface temperature of a star is near 10,000 kelvins, the hydrogen atoms are primed for eating visible photons. Spectra of these stars show discrete missing parts, or "spectral lines," where photons were absorbed by hydrogen's electron. The pattern of these absorption features in a star's spectrum is the *unique signature* of hydrogen.

Every element in the Periodic Table of the Elements and every molecule also has a unique signature, which can appear in other parts of the spectrum. Major branches of astrophysics depend on our ability to interpret these spectral features. From the spectral features of a star, we can deduce directly or indirectly its chemical composition, whether it is a binary system, whether it approaches or recedes from Earth, the speed with which it approaches or recedes from Earth, the surface temperature, the mass, the velocity of expansion or contraction (if it pulsates by changing size), the rotation rate, its approximate size, and the strength of its surface gravity.

Ultraviolet

Ultraviolet light is composed of photons with higher energy and shorter wavelengths than visible light. Many insects that fly in the evening hours have their visual range shifted from that of humans. They have trouble detecting red light, but they are quite sensitive to violet and ultraviolet. This is why red-light "bug bulbs" over your outdoor dinner table will not attract bugs. It is also why the bug electrocutors (available at

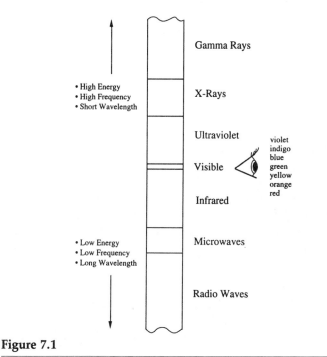

Figure 7.1

your local hardware store) have violet bulbs in them. The innocent, unsuspecting bugs fly toward the violet light, make contact with a cleverly placed electrified wire mesh, and are promptly zapped to death.

In the cosmic realm, we find high-mass stars whose surface temperature is hot enough for their energy output to peak in the ultraviolet. Ultraviolet telescopes spot these stars quite readily when interstellar dust does not obscure them. For example, a red star and a blue star might have the same brightness in the yellow part of the spectrum. But the blue star is likely to be many tens of times brighter in the ultraviolet.

Other objects that emit ultraviolet photons include close binary star systems that transfer their gaseous material, novae, supernovae, starburst galaxies, and quasars. In short, ultraviolet astronomers tend to specialize in exotic and energetic phenomena.

Ultraviolet from the Sun is mostly blocked by atmospheric ozone. The small fraction of ultraviolet photons that reach Earth's surface is enough to provide sun worshipers with body tans—and, of course, skin cancer.

X-Rays

While ultraviolet photons are only "skin deep," X-ray photons go right to the bone. Excess exposure to X-rays can induce assorted cancers of your internal organs. This is why the X-ray technician at your doctor's office promptly leaves the room and closes the door while you are being X-rayed.

The phenomena in the universe that produce X-ray photons are even more energetic and exotic than those which produce ultraviolet photons. X-rays are also blocked by atmospheric ozone. Special orbiting satellites detect X-rays from the 5-million-degree tenuous corona of the outer solar atmosphere, the disks of friction-heated gas that spiral inward to the surface of a pulsar or a black hole, the hot intergalactic medium of galaxy superclusters, and of course supernovae and their remnants. (Actually, supernovae are such titanic and energetic explosions that they release a full range of photons for everybody.) There is also a smooth background flux of X-rays that is detected everywhere we look. It is rationally termed the "X-ray background." There is still no agreement in the research community about its origin.

The first Nobel Prize ever awarded in physics went to Wilhelm Konrad Röntgen in 1901 for his discovery of X-rays.

Gamma Rays

Our slide through the electromagnetic spectrum now takes us to the variety of light with the shortest wavelength and highest energy of all types. These are the deadly gamma ray photons. Exposure to high-intensity gamma rays, when it does not result in death, nearly always triggers genetic defects. The famed *Hulk* of Marvel comic books is big, green, and ugly because of excessive exposure to gamma rays.

For a star's peak energy output to be in gamma rays requires a surface temperature of over a billion degrees. This is far beyond what we find for the surface temperature of any known star. The thermonuclear fusion of hydrogen into helium in stellar cores, however, releases copious quantities of X-ray and gamma ray photons. Unfortunately for these photons, there is no free trip out of the Sun. Their journey from the core to the surface involves about ten million years of getting absorbed and reemitted and scattered by the atoms and electrons along the way. The photon path is not unlike the random walk that drunk people exhibit upon exiting a bar. But unlike drunk people who can die in car wrecks before they reach home, the energy of the gamma photons emerges safely from the Sun. The form of this energy, however, was transformed from X-ray and gamma ray photons near the core to mostly lower-energy visible photons at the surface. A single gamma photon can spawn—at the cost of its own existence—more than 100,000 visible photons.

Gamma rays photons are almost entirely absorbed by Earth's atmosphere. Nearly all gamma ray astronomy occurs with the help of orbiting gamma ray detectors. The first of their kind were the American *Vela* satellites that were launched to monitor Earth-based nuclear explosions in the late 1960s. They promptly detected strong, short gamma ray bursts of cosmic origin. Strong bursts occurred and still occur about four to five times per year and come from all directions. The source of these bursts remains a mystery.

Emitters of gamma rays include supernovae and their remnants, the center of the Milky Way galaxy—and, of course, the Hulk when he is angry.

It is common for astronomers to devote entire careers to a single part of the electromagnetic spectrum. Indeed we are labeled that way. There are X-ray astronomers who have never looked through a visible light telescope, and there are radio astronomers who will never need to worry about lifting their telescopes into orbit. All astronomers know, however, that the deepest understanding of cosmic phenomena can only be achieved with the synthesis of knowledge that all photons in the family can provide. This is how we help to push forward the crests (and the troughs) of human understanding of the universe.

· 8 ·

Shapes of Radiation

Stars come in a staggering variety of sizes, luminosities, masses, temperatures, and densities, but they do not come in every color. Nobody has ever seen a lime green star, a chocolate brown star, or a star that was bubble-gum pink. A conspiracy of neurophysiology and astrophysical law mandates that stars come in only three basic colors: reddish, white, and bluish. Unfortunately, this fact is not revealed by simply looking up at the nighttime sky. As seen from Earth with the unaided eye, most visible stars are not bright enough to trigger the color-sensitive "cones" of the human retina. With the help of a telescope, however, the red, white, and blueness of stars can be quite striking. The John Philip Sousa tune that has the lyric, "Three cheers for the red, white, and blue," might well be adopted to honor stars in the universe.

☙

*T*here is much to explore among the photons of light that emerge from objects that glow. Of the several properties that a photon can have, the correlated triad of energy, frequency, and wavelength is paramount. Photons of high energy always have a high frequency and a short wavelength, while all photons of low energy have a small frequency and a long wavelength. It is not widely appreciated that humans have many more measurable properties than photons and other particles, which renders human behavior to be much more complicated than particle physics. A correlated triad of properties among humans might be height, weight, and the number of walking strides per block. Taller people, on average, weigh more than shorter people, and they generally take fewer steps to get to where they are going. Of these three properties, height is most fundamental because it does not fluctuate with eating habits, nor does it lengthen when you are in a hurry.

The Shape of Height

There are various ways that we can investigate the height of the world's humans. One is to line up everybody in size-place. (Incidentally, this line would be 2 million miles long as it wrapped about eighty times around Earth.) While this would make an interesting international activity, it would not be as informative as if you simply grouped people of similar height and counted them. You could begin by using 12-foot intervals, but then nearly everybody would be collected into the 0 to 12-foot category. A slightly better choice would be to use 3-foot intervals. Most adults would fall into

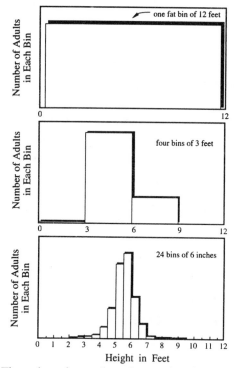

Figure 8.1. Three bar charts that show what happens when you change the bin size. Intelligently chosen bins will allow a bar chart to reveal more information about the sample.

the 3 to 6-foot category. The 6 to 9-foot category would have relatively few people although it would contain nearly all the players in the National Basketball Association.

A more sensible division might be six inches. This would give us twenty-four smaller bins in which to place people from 0 to 12 feet. If we look only at a sample of American adults, then we will discover that the 5-foot 6-inch to 6-foot bin holds the largest number. We would expect the bins on either side to have somewhat fewer people, and we would expect this downward trend to continue. Having carefully studied the heights of a large number of American adults, I can say with confidence that the four bins up to 2 feet and the four bins between 10 and 12 feet each contain very few people.

If you were to plot the number of people in each bin, then the graph would resemble what is known in scientific parlance as a "distribution function." A less oppressive and possibly more familiar term is a "histogram." If you prefer words that sound as though they were born together, then the most user-friendly term is probably the "bar chart," which has as a culinary cousin called the "pie chart." It is the peak and the shape of our bar chart (the relative number of people in each bin) that excites the person whose task it is to study height trends. For example, the bar chart for the height of American adults will look different from the bar chart for players in the NBA, and both will look different from the bar chart for female Olympic gymnasts. A comparison of these three populations reveals some obvious distinctions. The bin with the most NBA players is found at a larger height than that of the general population. And the bin with the most female gymnasts is found at a smaller height than that of the general population. Additionally, the range of occupied bins is broader for the general population than for either the basketball players or the gymnasts.

Of course, there is much more to learn from bar charts than the width and the location of the peak bin. A basic yet informative datum that can be extracted is the total number of people that are plotted. You get it when you simply add together the number of people in each bin. One can also ask whether the range of heights is symmetric on either side of the peak; or whether the average height equals the height that corresponds to the peak bin; or whether there are multiple peaks. Rest comfortably knowing there is an untold number of statistical tests that have been invented to analyze and compare the shapes of bar charts.

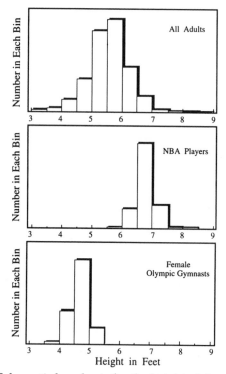

Figure 8.2. Schematic bar charts for the height of American adults, NBA players, and female Olympic gymnasts. Notice that the three charts peak at very different locations in height.

If your bar chart contains a very large number of people, then you have permission to use even skinnier bins (such as one-half inch). This will more finely subdivide the sample and ultimately force the bar chart to lose its steplike appearance in favor of what is best represented by a smooth curve. A convenient consequence is that smooth curves often lead to a mathematical model and then to physical insight.

The Shape of Light

In a discussion that is precisely analogous to drawing a bar chart for height, we may draw a bar chart for the energy of photons that emerge from an object that emits light. There is normally no shortage of photons for this exercise, so photon bar charts can be plotted with extremely narrow bins and are typically drawn as smooth curves. If we plot the photon energies emitted by a bright household incandescent (tungsten) light bulb, then we will reveal the component colors of what the eye detects and what the brain interprets to be white light. Some photons will be red, others will be yellow, green, and blue. If human eyes were sensitive to infrared light, then you would also notice infrared photons that emerge from the light bulb. In fact, most of the "wattage" of an incandescent bulb is wasted on these invisible photons.

The Sliding Peak

If you had unlimited control of the current that passes through the tungsten filament, you could do some amazing experiments. For now, however, a household wall dimmer-switch will do. As you increase the brightness, you should also realize that you are increasing the temperature of the filament as it receives more and more current. You may notice, along the way, that the bulb first glows with a deep reddish amber before it becomes white. If you were equipped with a super-dimmer and a tungsten filament that could carry a very high current (a quarter-million-watt light bulb, for example), then you could turn the dimmer until the light bulb glowed blue. As you adjusted the super-dimmer the tungsten would not only emit more and more photons, but the bin that contained the largest number of photons would shift from infrared to red and then across the spectrum of familiar colors (red, orange, yellow, green, blue, indigo, and violet), before it moved off to ultraviolet. This behavior leads to three basic scenarios for what the human eye sees:

1. Peak in the Infrared. At a temperature of 1,000 to 3,000 kelvins, the tungsten is cool. This assessment comes from the point of view of glowing objects, not from the point of view of your refrigerator. The interval of the bar chart that falls in the visible part of the spectrum is highest in the red and lowest in the blue-violet. When it is bright enough to excite the cones of the retina, this mixture of photons always appears reddish-orange to the human eye. These relatively cool temperatures are normally the hottest that an average person will ever encounter from everyday contact with household objects such as fireplace embers, "space" heaters, and heating elements of an electric stove—all of which is what started (and continues to feed) the rumor that red-hot is hot.

2. Peak in or near the Visible. At a temperature of 3,000 to 10,000 kelvins, the tungsten filament is much hotter than before (actually, the filament melts at 3,500 kelvins, but we will ignore that complication for now), and the bulb is much brighter. In fact, the bulb is brighter at every photon energy, including the red. If the peak photon bin falls anywhere near or between red and violet, then the filament will look white. For most glowing objects, including tungsten, the breadth of the bar chart near the peak is greater than the entire range of visible light. (You might even say that visible light has "bad breadth.")

A well-respected and often-worshipped example of a white glowing object is the Sun, whose surface temperature is nearly 6,000 kelvins. Our toasty neighbor is sometimes

Three Temperature Scales

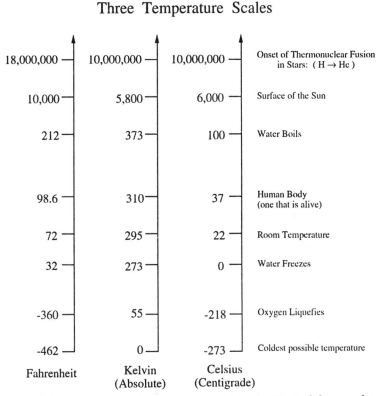

Fahrenheit	Kelvin (Absolute)	Celsius (Centigrade)	
18,000,000	10,000,000	10,000,000	Onset of Thermonuclear Fusion in Stars: (H → He)
10,000	5,800	6,000	Surface of the Sun
212	373	100	Water Boils
98.6	310	37	Human Body (one that is alive)
72	295	22	Room Temperature
32	273	0	Water Freezes
-360	55	-218	Oxygen Liquefies
-462	0	-273	Coldest possible temperature

Figure 8.3. The Fahrenheit scale is used in the United States, the Kelvin absolute scale (with degree units called *kelvins*) is used by the scientific community, and the Celsius scale is used by the rest of the inhabited civilized world.

claimed to be a yellow star. But this is only because the Sun's energy output peaks in the greenish-yellow part of the visible spectrum; with similar reasoning, you could, with equal aplomb, declare the Sun to be green. In the reality of your eyeball, however, your retina could testify that the color of the midday Sun has never resembled that of parsley,

Three Black Body Curves

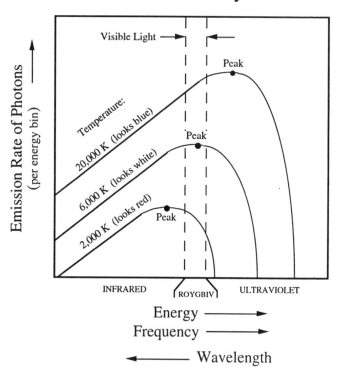

Figure 8.4 Black body curves for an object heated to three different temperatures. These schematic curves may be thought of as smoothed-out bar charts for the energy of photons that are emitted by glowing objects. The infrared, visible, and ultraviolet parts are noted along the bottom. The visible spectrum is further divided into ROYGBIV (red, orange, yellow, green, blue, indigo, and violet). Notice how the coolest curve passes through visible light higher in the red than in the violet, and the hottest curve intersects visible light higher in the violet than in the red. For intermediate temperatures, the curve is broad enough to contribute nearly equal quantities of red through violet, which will look white to the human eye. Also notice that the peak of each curve slides to higher energy (from infrared, through visible, to ultraviolet) as the temperature rises.

or of lemons, unless you live near the sulfurous outgassing of an active volcano. Indeed, there are enough visible photons emitted on either side of the greenish-yellow peak to excite equally your red, green, and blue retinal cones. It is this combined stimulus that allows the photon mixture from the Sun and from the tungsten filament to appear white. If the Sun were indeed yellow, then everything on Earth that normally looks white in daylight would look yellow—snow included. Although if you have ever actually seen yellow snow in the daytime, then it was probably near a fire hydrant.

For a star to look green to the eye it would have to emit *most* of its energy in the very narrow subset of the visible spectrum that we call green—there can be very little red-orange and blue-violet light to "contaminate" what you see. One must realize, however, that this would be a very unphysical thing for a star to do.

3. Peak in the Ultraviolet. At temperatures greater than 10,000 kelvins, the tungsten filament is extremely hot (the filament has, in fact, vaporized, but we will ignore that complication for now), and the bulb is very much brighter than before—not only to the eye but over all photon energies. Ultraviolet photons are invisible, but among the emitted visible photons the bin with the largest number is violet, while the bin with the least number is red. When bright enough to excite the cones of the retina, this mixture of photons always looks bluish-violet to the human eye, in spite of the fact that a blue-hot filament emits more red light than a red-hot filament. Perhaps astrophysicists and industrial welders are the only people on the planet who maintain a daily awareness that blue-hot is hotter than red-hot.

Wilhelm Wien, a turn-of-the-nineteenth-century physicist who won the Nobel Prize for physics in 1911, first formulated the relation that describes the slide of the peak photon bin as the temperature rises. This is now known as Wien's displacement law and can be written quite simply as

Surface Temperature ∝ Energy of Photons in Peak Bin

It reveals that the surface temperature of an object (tungsten, in the case of our filament) is directly proportional to the energy of the photons at the peak of the curve. The squiggle that resembles a goldfish is the international "proportional to" symbol. For example, if you increase the temperature of an object by a factor of three, then you can fully expect the energy of the peak bin to increase by a factor of three. Wien's law can similarly be written as

$$\text{Surface Temperature} \propto \frac{1}{\text{Wavelength of Photons in Peak Bin}}$$

which declares that the surface temperature of an object is inversely proportional to the wavelength of photons in the peak bin. This follows naturally when we remember that every photon can be identified with a particular wavelength that *decreases* with increasing energy.

The Totals

The total number of people in our bar charts could be obtained simply by adding up the contents of each bin for all heights. A similar procedure also allows us to compute the total emission rate for photons of all energies from each curve. For a proper comparison between objects of different size, all that we require is that the total emission rate be computed over the same-sized patch for each object. Fortunately, this has already been done for us. It is represented by what is known as the Stefan-Boltzmann law from the pioneering efforts of the nineteenth-century physicists Josef Stefan and Ludwig Boltzmann. The law reveals that for an object that maintains its temperature, the total emission rate of photons is proportional to the third power of the temperature.

Total Emission Rate of Photons $\propto$ (Temperature)^3

What makes this statement remarkable is that the total emission rate of photons from an object is unrelated to everything else about the object, such as its composition, its shape, or even its state of matter (solid, liquid, gas, plasma)—which is why our tungsten superfilament worked for our earlier example, even though it melted and then vaporized.

Sometimes (in fact, most times) it is easier to measure the total emission rate of *energy* in photons rather than the total emission rate of photons themselves. When this is done, then the Stefan-Boltzmann law takes the form

Total Emission Rate of Energy in Photons $\propto$ (Temperature)^4

With temperature raised to the fourth power, it boggles the senses to realize how much extra energy is emitted with a small change in temperature. For example, a blue star that is the same size as a red star, yet five times as hot on its surface, will radiate at a rate that is $5 \times 5 \times 5 \times 5 = 625$ times larger than that of the red star.

What stars and tungsten filaments have in common is that the peak and total energy of their emitted photons can each be described by the same basic laws. But these basic laws emerged from a simpler concept of an ideal black absorber.

The Shape

The term "black body" is simultaneously a nomer and a misnomer. It is well known that perfectly black objects absorb all photons that hit their surface. But it is not frequently discussed at dinner parties that the absorption of photons cannot continue without various thermal consequences—black bodies also emit photons. If more photons are absorbed than emitted, then the object's temperature will rise, and its total output of photons will increase until the emission rate equals the absorption rate. Anybody who wears a bathing suit and unwittingly sits on the upholstered black vinyl seat of a car with closed windows that has been parked several hours in the midday summer sun will testify nicely to this fact. Even if you live in the Arctic and have never conducted this butt-burning science experiment, then trust me: the car seat will be *much hotter* than the outside air temperature—indeed, it will be aglow (though invisibly) with infrared photons.

What makes an even better absorber than a black vinyl car seat is a small hole in an otherwise enclosed box that has a painted black interior. You can kiss good-bye all photons that enter the hole. They are likely to be absorbed—even those photons that manage to reflect off the interior walls a few times. We already learned from the Wien and Stefan-Boltzmann laws that the energy of the peak photon bin, and the total emission rate of photons, is only related to the temperature. An impressive property of this hole-in-the-box experiment is that no matter what quantity or assortment of photons you send into the hole, the box's cavity will reradiate a specific ensemble of photons whose detailed *shape* in a bar chart is also related only to the temperature of the cavity. The shape of such a bar chart is formally called a "black body spectrum," which is a powerful analytic tool that also applies, with excellent accuracy, to such diverse objects as our tungsten light bulbs, the gaseous surface of the Sun, and almost anything else that maintains its temperature as it glows (visibly or invisibly) from being heated—black vinyl car seats included.

If tungsten, the Sun, and black vinyl car seats do not make a complete enough list for you, then consider the biggest black body of them all: the cooled remains from the primeval fireball of the big bang. This omnipresent radiation field of the universe is extremely well-fitted by a black body spectrum whose temperature is a mere 3 kelvins, which is why a single temperature can be assigned to the entire universe. As noted in chapter 7, this is the famous "3 degree background" that peaks in the microwave part of the spectrum.

A physical understanding of the black body spectrum was first realized through the efforts of Max Planck, a German physicist and the 1918 Nobel Laureate in physics. On December 14, 1900, Planck presented a derivation of the Black Body Radiation law at a meeting of the German Physical Society in Berlin, which assumed that heated (and thus vibrating) matter emits photons with discrete and quantized energies. This new "quantum" concept provided one of the earliest hints that the new branch of physics, which grew to be known as quantum mechanics, was fast-approaching. So as not to slight Planck in favor of Wien, Stefan, and Boltzmann, I present Planck's radiation formula. It precisely describes the energy (shape) of black body radiation at every location on a plot, which heralded a new fundamental constant, h, that now bears his name, if not his initial:

$$E(\lambda, T) \; \propto \; \frac{1}{\lambda^5} \times \frac{1}{e^{(hc/\lambda kT)} - 1}$$

The classical "$1/\lambda^5$" part was known to Wien and others, but Planck's quantum adjustment of "$1/(e^{(hc/\lambda kT)} - 1)$" was brand new. The symbol E represents the energy emission rate in photons, λ is the Greek letter lambda that represents the photon wavelength, T is the temperature of the object, e is the exponential constant 2.71828 . . . , c is the speed of light, and k is a constant named after Boltzmann that allows the conversion of temperature units into energy units.

With some basic and some flashy calculus, you can actually derive Wien's law and the Stefan-Boltzmann law from Planck's radiation formula, but we will not do that here.

Reflections

There is no need to be upset that glowing, heated objects can only appear red, white, or blue. In the cool world of reflected light, one can be quite choosy about the colors of objects. When selecting paint for your house, or when selecting a flavor of Jell-O, there is (or seems to be) an uncountable number of colors from which to choose. Surface texture, pigments, and dyes all act selectively to reflect back a particular set of photons that you identify as a unique color. Parsley looks green only because the light that shines on it contains green photons (such as green light or white light) that are reflected back to you. If you illuminated parsley with light that contained no green photons (such as pure red or blue light), then the parsley would simply appear black.

Ordinary white surfaces look white simply because they reflect all visible photons. Approximately equal intensities of red, orange, green, blue, indigo, and violet light will always be perceived as white. This reflective fact also means that white objects will assume the color of the light that shines upon them. One night a few years ago, I was collecting data on the universe from a telescope dome whose interior was illuminated with deep red night-lights. When I took a lunch break around midnight, I spotted a soft-drink can that was leaning against one of the computers. The can occupied the next twenty minutes of my attention because it had positively no writing on it, not even a picture of a person, place, or thing. It just looked red. At the end of the night, when I finally turned on the white lights, I noticed that my midnight mystery was simply a Coca-Cola can with its familiar red-and-white design. Since red reflects red, and white reflects red, and the can was illuminated with a red light, then the *entire* can looked red with no written features to distinguish it.

While we are on the subject, it is possible to use the principles of the mysterious soft-drink can to one's own advantage. When I was a freshman in college, my physics professor allowed everybody to bring into the final exam a single $3\frac{1}{2} \times 5$–inch file card that could contain anything we chose to write on it—equations, notes from class, solved homework problems, and so forth. A friend of mine, frustrated by the tiny area upon which to put her helpful hints, filled both sides with tiny writing in blue ink. She then filled both sides again, writing in red ink directly over what was previously written. She entered the final exam with one red filter (that transmits only red light) and one blue filter (that transmits only blue light) with which she proceeded to extract all information that had been etched upon this otherwise unreadable puddle of red and blue scribble. While using the blue filter, the white file card turns blue, the blue writing remains blue, and the red ink looks black and readable. Similarly, while using the red filter, the white file card turns red, the red writing remains red, and the blue ink looks black and readable. I do not know what score my classmate earned on the final exam, but the color-coded crib sheet was brilliant.

If you were to mix red, orange, yellow, green, blue, indigo, and violet paints together you would get a color that closely resembles that of urban sewage. Clearly, colors of pigment do not add in the same way as do colors of light.

But this is not the only place where art deviates from physical reality. The "warm" paints are typically those one might choose for an artist's rendition of Hell; they include red, orange, and yellow. The "cool" paints are typically those one might choose to depict an Arctic igloo; they include violet and blue. You now know that among visibly glowing objects in the universe, the reverse of the artist's rendition is true—it is the coolest stars that are red, and the hottest stars that are blue. This cosmic awareness is so deeply etched in my mind that I do not know how much of my life I have wasted while staring at color-coded red and blue faucets as I attempted to deduce which knob would give me cold water. Out of concern for my repeated confusion, my dependable acquaintance once wrote

> On canvas with paint
> In the Artist's school
> It is red that is hot
> And blue that is cool.
>
> But in science we show
> As the heat gets higher
> That a star will glow red
> Like the coals of a fire.
>
> Raise the heat some more
> And what is in sight?
> It's no longer red
> It has turned bright white.
>
> Yet the hottest of all,
> Merlin says unto you,
> Is neither white nor red
> When the star has turned blue.
> *Merlin of Omniscia*[1]

1. From *Merlin's Tour of the Universe* (New York: Columbia University Press, 1989), p. 160.

Colorful poetry notwithstanding, if you ever see a motionless thirsty person who appears befuddled by the red and the blue spigots of an office water cooler, then say hello—it's probably me.

· 9 ·

The Hertzsprung-Russell Diagram

Not since Mendeleev's Periodic Table of the Elements has such a simple arrangement of data been so revealing as to how the universe is assembled. In the Hertzsprung-Russell diagram, the cosmic crucibles of fusion that we call stars are coherently organized in a way that allows astronomers to probe stellar evolution with dazzling detail. As important a discovery to astronomers as Darwin's theory of evolution is to biologists, the Hertzsprung-Russell diagram remains one of the twentieth century's greatest achievements in astrophysics.

☙

*T*he range of luminosities among stars in the universe is staggering. Not only is this range large—it spans a factor of hundreds of millions—but every measurable interval of luminosity between the extremes is represented somewhere by some star. The range of stellar surface temperatures is also impressive, although not as staggering—it spans about a factor of about fifty. What sort of questions must be asked to begin to understand this? Will the answer be complicated or simple? Stars also exhibit a mind-boggling range of sizes and densities. Can any organized understanding possibly emerge from stellar properties with such breadth? I once asked the Nobel Laureate Steven Weinberg a similar question about the state of particle physics: "In the interest of simplicity and elegance, shouldn't we be concerned that nature harbors such a zoo of subatomic particles?" to which he responded, "I don't care how many particles there are, just so long as they are described by only a few ideas."

As discussed in chapter 2, the greatest scientific ideas tend to be conceptually simple while they simultaneously create order out of confusion. In search of such a great idea, the Danish astronomer Ejnar Hertzsprung and the American astronomer Henry Norris Russell independently organized the luminosities and temperatures of stars in the diagram that bears their names, although it is often affectionately called the "H-R" Diagram. Actually, Hertzsprung and Russell could not measure a star's temperature, only its color. But temperature and color are physically related, so knowledge of the color is often just as useful. Their diagram serves a tremendously simple function. It displays the luminosity versus surface temperature for stars in the universe. When we consider the inherent range of these two stellar properties, one might expect the entire graphed plane to become filled with points—perhaps all combinations of luminosity and temperature are possible. Actually, such a result would be scientifically useless because it would not lay bare the hand of universal law. In one sweep of plotted data, however, Hertzsprung in 1905, and Russell in 1913, each determined that all combinations are *not* possible and that these stellar properties are clumped in regions on their diagram that reveal the existence of entire stellar families.

Stellar Families

As organized on the H-R Diagram, most stars in the universe, including the Sun, fall along a diagonal sequence from hot and luminous down to cool and dim. In the typical naming fashion among astronomers, this main sequence of stars is officially called the "main sequence." Another assembly appears at relatively high temperature yet has a very low luminosity. The region is cleanly delineated below the middle region of the main sequence. Two other regions are also relatively well defined. One set is a band of extremely luminous objects that spans a large range in temperature, and another set forms a collection of very luminous red objects above the main sequence to the middle-right of the diagram. *All other regions are relatively empty.*

That all stars assemble in families when plotted in a temperature-luminosity diagram is not unlike dishes that fit in a dishwasher. A sinkful of dinner dishes gets systematically sorted as you load them into the machine. All the silverware goes together in the silverware holder, all the drinking glasses assemble neatly on the upper rack, and all the plates form a tidy row on the bottom. Most people (I presume) take for granted the cosmic significance of this domestic task.

H-R Diagram

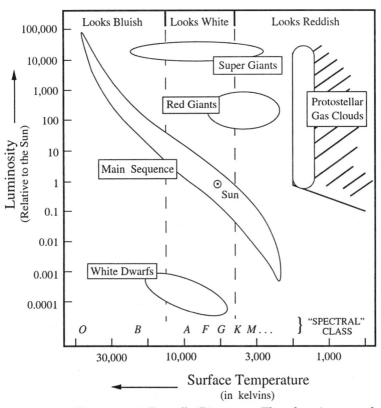

Figure 9.1. Hertzsprung-Russell Diagram. The locations and names of the principal stages of stellar evolution are noted. By tradition, the surface temperature increases to the left.

Consider that if no two items to be loaded in a dishwasher were alike, then there would be no pattern to reveal itself among the presized racks. This could happen, for example, if you had aliens from several different planets over for dinner, and they brought their own dishes.

It is always the task of a research scientist to find ways to look at data that will reveal trends or patterns. In this effort, Hertzsprung and Russell were triumphant.

Stellar Sizes

It is well known, from the study of radiating objects, that if you measure the rate that energy emerges from a chosen area on the surface of a hot star, then it will be much greater than the rate of energy that emerges from an equal-sized area on the surface of a cool star. This physical principle was expressed formally in chapter 8 in what is known as the Stefan-Boltzmann law. The law, which makes sense when you think about it, arms you with a veritable treasure map as you amble among the stellar clumps of the H-R Diagram. With no information other than an H-R Diagram in one pocket and the Stefan-Boltzmann law in the other, you can decipher the exact relative sizes of stars in the universe. For example, the blob of stars that lurks in the lower-middle of the H-R Diagram falls at the same temperature as many of the stars on the main sequence. But the stars of the lurking blob are about 10,000 times dimmer. The only way for this to be true is if the radiating area of each of these stars is 10,000 times smaller. Using simple formulae for the geometry of spheres (all stars are conveniently spherical), one discovers that our lurking stars are each about as small as Earth. Most of them being white, and all of them being dwarf-ish, astronomers figured that "white dwarf" would be an appropriate official name for them.

Somewhat above the "cool" region of the main sequence, we find a blob of stars whose luminosity ranges from 100 to 10,000 times that of the main sequence stars below it. The only way cool stars can become this luminous is if the surface area over which they radiate is 100 to 10,000 times larger. Most of them being red, and all of them being giant-size, astronomers figured that the name "red giant" might be appropriate.

Perched above the red giants is yet another blob of stars that are one hundred times more luminous than the red giants themselves. Stars in this group are found to be red, white, and blue. The coolest among them have the same temperature as the red giants and must therefore radiate with one hundred times the surface area. I do not mean to shout, but THESE STARS ARE ENORMOUS. If you put one of them at the location of the Sun (about 93 million miles from Earth), it would be large enough to engulf our entire planetary orbit—and, of course, it would vaporize Earth. Astronomers were left with no choice: being bigger than the "normal" giants, these super-big stars that compose our superluminous blob are simply called "supergiants," with an occasional reference to their color, as appropriate. The constellation Orion is fortunate enough to contain, as its two brightest stars, the red supergiant Betelgeuse (Alpha Orionis) appearing as Orion's right armpit, and the blue supergiant Rigel (Beta Orionis) appearing as Orion's left kneecap.

We just "sized-up" the principal regions of the H-R Diagram without ever traveling through space to see them up close. We succeeded, however, with the assistance of specially constructed scales. For convenience, and to save trees, the H-R Diagram is normally plotted with both its luminosity and temperature scales in logarithmic form. This simply means that the scale gets very big, very quickly, so that you can fit everything on one sheet of paper. In this case, the distance on the temperature scale between 300 and 1,000 kelvins is the same as that between 3,000 and 10,000 kelvins. And the distance on the luminosity scale between 1 and 100 is the same as the distance between 1,000 and 100,000. Without a logarithmic scale, our H-R Diagram would require a foldout insert that measured five miles high and one yard wide—to the disgust of my publisher.

In an attempt to classify the sizes of stars, the American astronomers William Wilson Morgan, Philip Childs Keenan, and Edith Kellman pioneered the use of "Luminosity Classes" for all stellar groups in the H-R Diagram with the 1943 publication of a major atlas of stellar spectra. The supergiants were labeled Class I, the red giants were labeled Class III, the entire main sequence was labeled Class V, and the degenerate white dwarfs were labeled Class VII. The remaining intermediate Roman numerals refer to stars that occupy underpopulated regions of the diagram. They are: Class II (the bright giants); Class IV (the subgiants); and Class VI (the sub-[main sequence]-dwarfs). As a main sequence star, the Sun is simply luminosity Class V.

But before anybody knew about luminosity classes, before anybody knew the relative sizes of stars in the universe, before anybody knew how to measure the temperatures of stars, indeed almost before anybody knew much of anything—there were spectral classes.

Spectral Classes

A star's spectrum, if you photographed it with color film, would typically show the colors of the rainbow. Upon very close view, however, you would also notice many places where slices of the spectrum are missing. These slices, or spectral lines, are where the atoms in a star's atmosphere absorbed photons of their choice. The element hydrogen, which comprises over 90 percent of the Sun's atoms, has a photon-absorbing signature that shows up nicely in the visible spectra of most stars. Between 1918 and 1924, Annie Jump Cannon, the Curator of Astronomical Photographs at the Harvard College Observatory, with unmatched visual acumen used the prominence of this hydrogen absorption to classify the spectra of over 225,000 stars for the *Henry Draper Catalogue*. Through her continued efforts, subsequent catalogue supplements brought the total to over 350,000. These spectral classes were organized by letter, beginning with A:

A B C D E F G H I J K L M N O P Q R S
<u>prominent hydrogen feature   weak hydrogen feature</u>

With the discovery of quantum mechanics in the 1920s, a theoretical basis emerged for the variation in spectral features. What was previously a descriptive "botanical" classification now found a conceptual anchor: hydrogen atoms that tend to absorb emerging photons share the same worldview as that expressed in the fairy tale *Goldilocks and the Three Bears*. To assist your memory of the story, here is my synopsis of the relevant part:

> Hungry and tired, Goldilocks, upon stumbling across the Three Bears' empty cottage in the woods, found three bowls of porridge (whatever that is). The porridge in one bowl was too hot, that in another bowl was too cold. But the third bowl of porridge was "just right," so she ate it. The Three Bears came home later.

Quantum mechanics showed that hydrogen atoms in stellar atmospheres are as choosy about photons as Goldilocks was with other people's porridge. It is possible for a star to display a weak hydrogen feature in the visible spectrum from being too hot *or* from being too cold. This humble discovery was the birth of modern astrophysics. The hydrogen signature is most prevalent when a star's surface temperature is "just right," which is about 10,000 kelvins. While we presume this to be somewhat hotter than Goldilocks' preferred porridge, it is neither the hottest nor the coldest of stellar atmospheres. Armed with this physical understanding of the process, it became more sensible to catalogue stars as a sequence of surface temperature. Some classifications were dropped, and others were consolidated. What remains are some leftover letters which, when arranged from hottest to coldest, look like

$$\underline{\text{O} \quad \text{B} \quad \text{A} \quad \text{F} \quad \text{G} \quad \text{K} \quad \text{M}}$$
$$\textit{hottest} \qquad\qquad\qquad \textit{coolest}$$

Three other letters, R, N, and S, are not formally included in the temperature sequence because they identify subsets of K and M stars that display peculiar features. The letters O and M are far from A in the alphabet. As already noted, this implies weak hydrogen features as a spectral class—notice that both O and M fall at the extremes of the temperature scale.

This new and improved sequence of letters is world famous because it has become the darling of mnemonic writers. Henry Norris Russell unwittingly started the mnemonic movement after he suggested, "Oh Be A Fine Girl, Kiss Me." Those astronomers who did not want to alienate the R, N, and S categories soon added to the original: "Oh Be A Fine Girl, Kiss Me Right Now; Smack!" It is not clear whether the *Smack!* was intended to be the sound of a kiss or the sound of a male astronomer getting punched in the nose for making such a request, but the mnemonic was widely celebrated. As astronomy got more complicated in the latter part of the twentieth century, frightened college freshmen devised, "Oh Boy, [my] Astronomy Final's Gonna Kill Me." And a recent mnemonic, donated by a frustrated male freshman, gets the award for licentiousness: "Obvious Bulges Always Frighten Girls Kissing Me." You will probably never see that one in a textbook.

One does not automatically obtain a temperature from the presence or absence of hydrogen features. Typically, a temperature is derived from a star's color, which can be quantified with impressive precision if you measure a star's emission through specially chosen sets of colored filters. The proportions in which the different colors appear, when compared with what you would expect from a perfect black body, allows an accurate assessment of a star's surface temperature. This extra precision permitted each lettered spectral class to be reliably subdivided into ten parts. The spectral class "G," for example, runs from 0 to 9 before the onset of class "K." Our beloved Sun finds itself at spectral class G, subclass 2. When combined with its luminosity class, the Sun, in shorthand, becomes a G2V star. This corresponds to a main sequence "dwarf" with a surface temperature of 5,800 kelvins. At 10,000 kelvins, the bright main sequence star Vega (Alpha Lyrae), in the constellation Lyra, is designated A0V. And at 27,000 kelvins, Alkaid (Eta Ursa Majoris), the star at the tip of the handle in the Big Dipper, checks in at B3V.

Stellar Evolution

A naive view of the stellar groups in the composite H-R Diagram may lead you to suspect that there are four types of stars that are born and die as you see them. Or maybe that stars are born as hot O stars, and when they age, they cool, as they slide down the main sequence to become an M star. Or maybe stars begin as bloated supergiants and then collapse in sudden jumps to become giants, main sequence stars, and then white dwarfs. Or maybe stars are born as white dwarfs and somehow get swollen over time to become giants. Or maybe none of the above.

An equivalent glimpse of the general human population would raise similar questions to our alien dinner guests (the ones who brought the mismatched dishes). Are people born large, and over time shrink to become babies? Are people born with dark skin that lightens over time? Are people born with wrinkled skin and then expand on the inside when they get older? Are people born in the dirt, wrapped in wooden boxes, and then pop up out of the ground fully grown? Some (or all) of these questions sound completely stupid. But it is only because you, a human being, have the enviable status of already knowing the correct answer. Scientific pursuit at the perimeter of discovery is not always graced with such benefits.

Unquestionably, the greatest utility of an H-R Diagram is what it can reveal about a cluster of stars that were all born at about the same time. Astronomers officially call these clusters of stars "star clusters." The H-R Diagram for a star cluster is a mere snapshot in time—and none of us will live long enough to make an interesting motion picture of it. Actually, a star cluster that had been filmed continuously since the dawn of Homo sapiens would not show much either. The next best thing we can do is to assemble as many H-R Diagrams of star clusters as we can find. Indeed, this exercise tells a remarkable story of stellar evolution. When combined with mathematical models of stellar structure, the life cycle of a star is revealed.

When a protostellar gas cloud collapses under its own force of gravity, then the temperature of its core rises. In this stage the cloud is large, and most of its luminosity is emitted in the infrared. If you were to plot this unborn star on the H-R Diagram, it would incubate in the region of high luminosity and very cool temperature. If our nascent protostar has sufficient mass, then the core temperature will rise high enough (10 million kelvins) to spark thermonuclear fusion. Not only is hydrogen transmutated into helium, but matter is converted to energy. When this happens . . . a star is born.

The highest-luminosity stars are born with high mass (up to one hundred times the mass of the Sun) and are extremely rare in any volume of space. Their impressive luminosity, however, allows plenty of them to be noticed out to great distances. The lowest-luminosity stars are born with low mass (as little as one-tenth the mass of the Sun). They are the most common variety of stars in space—although it would be hard to convince yourself of this from a list of the brightest stars of the nighttime sky. Typically only 1/1000th of the birth mass of a star is consumed through thermonuclear fusion over its entire life on the main sequence, which is a luminous testament to the power of nuclear fusion.

While hydrogen is being converted to helium in its core, the luminosity and temperature of a star will always place it within the main sequence of the H-R Diagram. So when the moment is right, the surface temperature of our infrared-emitting protostar will rise as it vaults across the diagram to land on the main sequence. A star spends 90 percent of its life there, so it is no coincidence that 90 percent of all stars are found on the main sequence. But what does it do next?

A montage of cluster diagrams shows that the high-mass, high-luminosity stars are terribly short-lived. They evolve from the main sequence in less than ten million years to become supergiants, which last only for a cosmic moment or two—about one hundred thousand years. The rarity of high-mass stars is only partly related to their short life span. In a freshly born cluster, the low-mass stars can outnumber the high-mass stars by thousands to one. The revenge of the high-mass stars, however, is realized through their blazing luminosity. In any new cluster, it is always the handful of high-mass stars that completely dominates the total light. In the end, those supergiants that do not collapse and disappear to form a black hole will die a violent death as a supernova explosion. All factors considered, if I were a high-

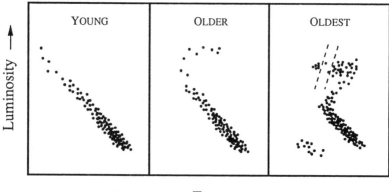

Figure 9.2. Hertzsprung-Russell diagram of a star cluster at three evolutionary stages. The stars of a young cluster, when plotted in the luminosity and temperature axis of the H-R Diagram, will lie along the main sequence. When somewhat older, the highest mass stars (the most luminous) will peel away from the main sequence to become the supergiants. At old age, the main sequence has peeled some more. Intermediate mass stars are now becoming red giants, while stars that had already been red giants have died to become the underlumious white dwarfs. For the oldest cluster, the underpopulated "instability strip" is drawn (dotted lines) where one is likely to find periodic variable stars. For all known clusters, there are always more stars born to the lower main sequence (with low mass) than to the upper main sequence (with high mass).

mass star, I would rather die as a supernova because your exploded corpse entertains astronomers on Earth *and* you remain in the observable universe.

Higher-mass stars have more mass (fuel) in their core, but they exhaust it inexcusably fast, like the oversized, gas-guzzling American cars of the 1950s and 1960s. These automotive monsters creations had big fuel tanks, but their gas mileage was so poor that you needed the big tank just to drive yourself to the next gas station.

After several billion years, all the stars from the high-mass end of the main sequence have become giants. With a life expectancy of several billion years, intermediate-mass stars are now observed to peel from the main sequence. This "a-peeling" transition happens relatively quickly, but if a cluster has many stars (thousands) then you will almost always catch one at every step of the transformation. As was true for the supergiants, the red giant phase lasts only about 10 percent of a star's life span. For the intermediate-mass stars, this amounts to several hundred million years.

When all usable nuclear fuel is exhausted, the red giants gently lose their atmospheric envelope to interstellar space and lay bare a hot, dense, degenerate core of compacted protons and electrons that is otherwise known as a white dwarf. The lost envelope becomes a planetary nebula. White dwarfs (some of which are hot enough to look blue) have no energy source. They are dead. All they can do is sit there in space and cool off. As their temperature drops, so does their luminosity. The double drop is what takes them diagonally down the H-R Diagram, like a sinking dirigible. They ulti-mately become red, and then invisible, as their luminosity slides into optical oblivion. Again, we do not actually watch this happen. We do not live long enough. Our conclu-sions are drawn from carefully compared H-R Diagrams of clusters and from mathematical models of dense matter as it cools.

After several more billion years, we have exceeded the age of the universe, for which there is clearly no example of such a cluster. But if we could find one, then we would expect the lowest-mass stars to still be chugging along. They have the least amount of fuel, yet they use it very, very, very slowly. Indeed, low-mass stars are the envy of the automotive industry, with a fuel efficiency that grants them a life expectancy that is hundreds of times greater than the age of the universe. One might call that eternal life—but I'd still rather be a supernova.

Stellar Distances

A glowing light bulb, with a particular wattage stamped on top, will remain that wattage no matter where in the universe you put it, as long as you keep the lamp plugged in. The bulb will, however, appear dimmer or brighter depending on its distance from you. The Sun, as a light bulb, would have 400 yottawatts (4×10^{26} watts) stamped on top. But is Polaris (the North Star) only the forty-ninth brightest star in the nighttime sky because it is extremely luminous and far away, or because it is underluminous and nearby? You get the brightness for free; it is easily measured with a light meter. But the luminosity—the wattage stamped on top—remains uncertain. This converts directly to an uncertainty in distance, which constitutes one of the greatest intellectual obstacles that an astronomer faces in the line of duty. Let it be known that there is surely a Nobel Prize that awaits the person who devises a way to measure directly the distances to everything in the universe.

One method to infer luminosity, which allows you to bypass the distance dilemma and which is reasonably accurate for main sequence stars, requires that you first estimate the star's temperature. One can straightforwardly obtain this from the *measured* color, which will be more precise than the color estimated by simply looking at the star with the unaided eye. Next, simply locate this temperature on the temperature axis of the H-R Diagram and slide upward until you intersect the main sequence. From that spot, all you need do is slide sideways and read the appropriate tick-mark from the luminosity axis. The diagonal property of the main sequence ensures that every temperature is accompanied by a unique luminosity, although the inherent thickness of the main sequence introduces some unavoidable uncertainty. The distance now follows readily from a simple and widely invoked formula that uses the measured brightness and the newly found luminosity:

$$Brightness \propto \frac{Luminosity}{(Distance)^2}$$

This basic formula can be used for stars in the galaxy and for galaxies in the universe. It will even work with your well-traveled light bulb.

An improved distance indicator would somehow use the giants, because you can detect them at greater distances than main sequence stars. Fortunately for astronomers, there are specific combinations of luminosity and temperature among the giants that are somewhat upsetting to their structure. This region is dubbed the "instability strip" because it is where a star becomes—you guessed it—unstable.

Most of these stars pulsate as their outer gaseous envelopes expand and contract. Two famous varieties of unstable giants are the "RR Lyrae" variables and the "Cepheid" variables. Their instability leads to predictable and periodic fluctuations in brightness, which correlates well with their true (average) luminosity. Nearly every star passes through the instability strip at least once in its life. In spite of this, variable stars remain relatively uncommon because they do not spend much time being unstable. Similarly, a snapshot taken of you at a random time of day will likely show you at work, or asleep, but not in the bathroom. This is true even though you presumably go the bathroom at least once a day; you simply do not spend much time there (unless, of course, you have gastrointestinal problems that prevent you from leaving). Such is the way with the rapidly evolving stars in the instability strip of the H-R diagram.

The RR Lyrae and Cepheid variables together represent a key step in the ladder of methods that allows us to derive distances within the Milky Way and beyond.

Your Ex-Lover

We have chosen the scale of our H-R Diagram to display the major categories of known stars, inclusive of the unborn and the postdead. But in principle, anything that radiates can have its temperature and luminosity plotted. You could even plot your ex-lover. This person probably wasn't so hot—about 310 kelvins sounds correct. This temperature falls off-scale to the right on the H-R Diagram. Your ex-lover probably wasn't so bright, either; about 1/4,000,000,000,000,000,000,000,000,000th the luminosity of the Sun is likely to be correct[1]. This luminosity falls far off-scale downward on the H-R Diagram. Actually, regardless of how you feel about your ex-spouse, most other (living) people on Earth would fall near your ex-spouse in the plane of the H-R Diagram, including you.

The Hertzsprung-Russell Diagram now serves as the bedrock of our understanding of stellar evolution. It is an indispensable tool to nearly every professional astronomer. And since the most basic constituents of the universe are stars, one can make a strong claim that the progress of twentieth-century astronomy would have been severely retarded in the absence Hertzsprung's and Russell's a-peeling contribution.

---

1. This is "one four-septillionth" of the Sun's luminosity, which is about equal to the luminosity of a 100-watt light bulb. The absolute temperature of 310 kelvins equals the average human body temperature of 98.6° Fahrenheit.

· *10* ·

Highlights from the Periodic Table of the Elements

Second only to the multiplication table, or perhaps King Arthur's Round Table, the Periodic Table of the Elements is the most famous table in the world.

❧

*T*he Periodic Table of the Elements, you may remember, is that mysterious chart of boxes with cryptic letters in them that hung in the front of your high school chemistry classroom. The Table is packed with information that excites the chemist, intrigues the lay person, and bores the poet. It contains the ninety-four elements found naturally in the universe and fifteen elements that are not.

You may also remember the subatomic particle trinity—electrons, protons, and neutrons. Atoms normally contain protons and neutrons in their nucleus surrounded by clouds of electrons in quantum orbits. The Periodic Table displays a sequence of atoms whose identities are defined by the number of protons they contain. By this simple scheme, the table "starts" with atomic number 1, the element with one proton in its nucleus. And it ends (for now) with atomic number 109, the element with one hundred and nine protons in its nucleus. The Periodic Table of the Elements is complete from 1 to 109. Over this interval, we will not one day discover an element that has to be inserted somewhere between two consecutive elements. The table contains all known elements that have ever existed in the universe or in our laboratories.

You might ask, "If the Periodic Table is just one long sequence, then why isn't it displayed horizontally in the classroom the way the alphabet is displayed in elementary school?" The ordered beauty of the Periodic Table is found in its grouping of elements with similar chemical properties. In a normal, undisturbed atom, we find that as the number of protons rises from 1 to 109, so does the number of electrons. They surround the nucleus in larger and larger patterns of clouds that contain up to 2, 8, 18, and 32 electrons.

The exact number of electrons per cloud along with the cloud's size and orientation is determined from the quantum mechanics branch of twentieth-century physics (no relation to auto mechanics). Notice that rows one through six down of the table contain 2, 8, 8, 18, 18, and 32 elements. Row seven still awaits the discovery of atomic numbers 110 through 118 for it to be a complete row of 32 elements.

The chemical properties of an element (such as the list of elements with which it can bond to form a molecule) are determined primarily by the behavior of its outermost electrons. Elements found vertically in columns have their outer electrons similarly configured and thus tend to form compounds with the same elements. Since these chemical properties repeat row after row, we find that what might have been named "The Element Table" has earned the more lofty title "The Periodic Table of the Elements."

Since an element is defined by the number of protons in its nucleus, we conclude that neutrons can come and go without altering the element's identity. In nature (this includes Earth and the rest of the universe) there is a well-researched number of neutrons that the nucleus of each element is most likely to have. For some elements, however, there are isotopes where their neutron count renders them unstable. In response, such an element will transform into the same element, but with fewer neutrons, or in other cases it will decay into a lighter, completely different element. Each time such a transmutation occurs, energy is released that can, at times, be enormous. This is not magic. These are elements following well-understood laws of nuclear physics. The versions of elements that readily decay are termed "unstable," although some elements are unstable no matter how many neutrons you give them. These are the famous "radioactive" elements such as radon, uranium, and plutonium.

Periodic Table of the Elements

Legend:
6 **C** Carbon — ATOMIC NUMBER / SYMBOL / NAME

1 H Hydrogen																		2 He Helium
3 Li Lithium	4 Be Beryllium											5 B Boron	6 C Carbon	7 N Nitrogen	8 O Oxygen	9 F Fluorine	10 Ne Neon	
11 Na Sodium	12 Mg Magnesium											13 Al Aluminum	14 Si Silicon	15 P Phosphorus	16 S Sulfur	17 Cl Chlorine	18 Ar Argon	
19 K Potassium	20 Ca Calcium	21 Sc Scandium	22 Ti Titanium	23 V Vanadium	24 Cr Cromium	25 Mn Manganese	26 Fe Iron	27 Co Cobalt	28 Ni Nickel	29 Cu Copper	30 Zn Zinc	31 Ga Gallium	32 Ge Germanium	33 As Arsenic	34 Se Selenium	35 Br Bromine	36 Kr Krypton	
37 Rb Rubidium	38 Sr Strontium	39 Y Yttrium	40 Zr Zirconium	41 Nb Niobium	42 Mo Molybdenum	43 Tc Technicium	44 Ru Ruthenium	45 Rh Rhodium	46 Pd Palladium	47 Ag Silver	48 Cd Cadmium	49 In Indium	50 Sn Tin	51 Sb Antimony	52 Te Tellurium	53 I Iodine	54 Xe Xenon	
55 Cs Cesium	56 Ba Barium	72 Hf Hafnium	73 Ta Tantalum	74 W Tungsten	75 Re Rhenium	76 Os Osmium	77 Ir Iridium	78 Pt Platinum	79 Au Gold	80 Hg Mercury	81 Tl Thalium	82 Pb Lead	83 Bi Bismuth	84 Po Polonium	85 At Astatine	86 Rn Radon		
87 Fr Francium	88 Ra Radium	104 Unq Unnilquadium	105 Unp Unnilpentium	106 Unh Unnilhexium	107 Uns Unnilseptium	108 Uno Unniloctium	109 Unn Unnilnonium	110 ?	111 ?									

57 La Lanthanium	58 Ce Cerium	59 Pr Praseodymium	60 Nd Neodymium	61 Pm Promethium	62 Sm Samarium	63 Eu Europium	64 Gd Gadolinium	65 Tb Terbium	66 Dy Disprosium	67 Ho Holmium	68 Er Erbium	69 Tm Thulium	70 Yb Ytterbium	71 Lu Lutetium
89 Ac Actinium	90 Th Thorium	91 Pa Protactinium	92 U Uranium	93 Np Neptunium	94 Pu Plutonium	95 Am Americium	96 Cm Curium	97 Bk Berkelium	98 Cf Californium	99 Es Einsteinium	100 Fm Fermium	101 Md Mendelevium	102 No Nobelium	103 Lr Lawrencium

Of the 109 elements, only twenty-eight were known in the year 1800. In 1814 the Swedish chemist J. J. Berzelius introduced the "initial letter" scheme by which element names are abbreviated. By 1850 another thirty-one had been discovered, which prompted worldwide interest in a means by which the elements could be ordered. The basic form of the Periodic Table as we know it today was first published in 1869 by the Soviet chemist Dimitri Ivanovich Mendeleev. Of course, there were some blank spaces, but by 1925 almost all the naturally occurring elements had been discovered.

Element names are normally assigned by their discoverers. These names form a rich collection of references to individual scientists, mythology, astrology, alchemy, geography, and chemical history. Most chemical symbols are derived from Latin, Greek, and German, which in nearly all cases resembles the common English usage. Among these 109 elements are some that deserve special attention here owing to their unusual properties, their relevance to human existence, or their importance in astronomy.

Let's meet these members of this not-so-exclusive table.

Hydrogen ($_1$H). With only one proton in its nucleus, hydrogen is the lightest of all elements. It is the only element that, in its most common form, contains no neutrons. In other words, if you ever see a proton all by itself, it's just the nucleus of a hydrogen atom. Over 90 percent of all the atoms in the universe and in the Sun are hydrogen. It should be no surprise that it is also the most abundant element in life on Earth.

In our dynamic universe, electrons come and go. At cooler temperatures it is normal for atoms to contain their full complement of electrons (the same number as protons) and are thus declared neutral. Under higher temperatures, an atom can lose one or more of its electrons and thus become "ionized." Ionized hydrogen would, of course, be found inside stars (in the form of hydrogen plasma), but it is also commonly found in gas clouds that are in the vicinity of blue-hot high-luminosity stars whose high-energy radiation kicks loose the lone electron.

"Neutral" hydrogen is probably the most common form of hydrogen in interstellar space. As noted in chapter 7, radio photons are occasionally emitted by neutral hydrogen atoms at a wavelength of 21 centimeters. These precious photons allow us to detect and map the gas content in gas-rich spiral and irregular galaxies.

In the coldest and densest of the gas clouds you are likely to find hydrogen chemically bound to itself to form "molecular hydrogen." These clouds of molecules are the homes of star formation that is soon to begin.

On Earth, hydrogen gas is isolated easily from the water molecule. It is lighter than air, which made it a popular filler for the early generations of blimps. But hydrogen is also extremely flammable. On May 6, 1937, the *Hindenburg*, a hydrogen-filled German dirigible (a blimp with ribbing), caught fire over Lakehurst, New Jersey. The heavier-than-air gondola could no longer enjoy the buoyancy of an attached blimp. Thirty-five died out of the ninety-seven passengers and crew. Hydrogen is no longer used to fill blimps.

Hydrogen under pressure exhibits properties that render it indistinguishable from a metal. This "metallic" form of hydrogen was first produced in 1972 at the Lawrence Livermore Laboratories in California. It is believed to be the primary component of Jupiter's core and is considered to be the source of Jupiter's tremendous magnetic field.

Because hydrogen is the simplest element, it forms the natural starting point for thermonuclear fusion. At a mere 20 million degrees Fahrenheit ("thermo"), hydrogen nuclei ("nuclear") will combine ("fusion") to form helium, the next heaviest element. The core of the Sun does this to about 100,000,000,000,000,000,000,000,000,000,000,000,000 hydrogen atoms (10^{38}) per second. This mass (con)fusion releases energy that supports the gaseous Sun against gravitational collapse and, of course, bathes Earth with life-sustaining light. In a remarkable duplication of this hydrogen fusion, humans have created life-annihilating hydrogen bombs that are the bulwark of today's nuclear arsenals.

Helium ($_2$He). It has been known to astronomers for nearly one and a half centuries that the presence of elements can be inferred from their telltale features in the spectrum of certain stars. In 1868 there was an element discovered in the spectrum of the Sun which nobody could readily identify from Earth-based experiments. This new element was named "helium" from *helios*, the Greek work for Sun. On Earth, helium was discovered twenty-seven years later.

Helium is the second lightest element, with two protons in its nucleus. It is a distant second in the universe's abundance list. No more than 8 percent of all atoms are helium. The atomic physics of the early, hot big bang is expected to have produced about 6 percent helium. The extra 2 percent of the helium was forged in the cores of stars over the last ten billion years of stellar evolution. Like hydrogen, helium can fuse to form heavier elements. But once a star has exhausted its core hydrogen supply (leaving helium as the residue), it may find that it is not massive enough to provide a high enough core temperature to fuse helium. Such a star will become a red giant and eventually lose its tenuous outer shells, thus exposing a small, hot helium core. This is the helium white dwarf.

Helium is lighter than air, and it is inert. Since it does not react chemically to anything, it is nonflammable. Helium became the immediate replacement gas for all blimps and dirigibles after the 1937 *Hindenburg* disaster. Life on Earth has no use for inert elements—the human body is 0.00 percent helium. But if you ever want to sound like a Munchkin from *The Wizard of Oz*, just inhale some helium and say hello to somebody. The fact that helium is inert means it won't harm you (unless that is the only gas you ever choose to breathe). And the fact that it is lighter (less dense) than air means your vocal chords will vibrate such that you produce a higher pitched sound. Incidentally, your voice would be even higher pitched if you inhaled some hydrogen, but if somebody lit a match near your mouth you would probably explode.

Lithium ($_3$Li). Lithium is a fragile element that is not favored in the thermonuclear fusion sequence. A little bit was produced in the big bang, but only moderately high temperatures and a spare proton are required to break apart lithium's three protons to form two helium nuclei. For this reason the cosmic abundance of lithium puts a strong constraint on the details of the high-temperature physics of the early universe.

Lithium on Earth is a very soft metal—you can cut it with a butter knife—that reacts explosively with water or water vapor. You have probably never seen it outside of a chemistry lab. In small doses lithium has a tranquilizing effect on the nervous system, which is why it is often prescribed in the treatment of manic-depressive illness.

Skipping a few elements brings us to . . .

Carbon ($_6$C). Carbon is a "sticky" element in the sense that it bonds strongly with many different elements (including itself), and it can be bound in many different ways. Some familiar forms of pure carbon include graphite (commonly found in "lead" pencils and charcoal briquettes), diamond, and the jumbo molecule buckminsterfullerine (or "bucky-ball")—named in honor of R. Buckminster Fuller for the similarity between this molecular form of carbon and Fuller's famous geodesic domes.

Life on Earth owes its diversity to the chemical properties of this single element. In short, life is carbon-based. You can prove this to yourself. Take any life form at all (which includes all the food you eat) and leave it in the oven too long. The heat will eventually break the molecular bonds and expose the black charred carbon remains. If you do this to a rock or a wrench, they will get hot, but they won't turn black. Carbon atoms comprise one-fourth of all the atoms in the universe that are not hydrogen or helium. It is an excellent bet that if life exists elsewhere in the universe, then it is carbon-based as well.

Carbon is created quite readily in high-mass stars that have converted their core supply of hydrogen into helium. The helium, when brought to over two hundred million degrees Fahrenheit, will form the very simple reaction

$$_2He + {}_2He + {}_2He = {}_6C + Energy$$

You will notice that the three helium atoms provide the required total of six protons to make carbon. Stars that undergo this reaction are in their red giant phase. A lower mass subset of these red giants will stop their thermonuclear fusion with carbon in their core. In time, the star's red outer shell will float off into space, laying bare its hot, dense core of carbon. This is a carbon white dwarf. White dwarfs do not make their own energy, so they cool continuously as they radiate into space. Carbon white dwarfs are affectionately called interstellar "charcoal briquettes" by some researchers in the field.

Nitrogen ($_7$N). It may be a surprise to some people that the air we breathe is composed of mostly nitrogen (about 78 percent). Nitrogen combines easily with three hydrogen atoms to make household ammonia (NH_3). Ammonia, as well as many other nitrogen and carbon-based molecules such as acetylene, formaldehyde, hydrogen cyanide, and methanol, are detected in cold interstellar gas clouds. These molecules tend to emit microwaves, which we detect on Earth with special microwave telescopes that look very much like cable television satellite dishes. Thus we have a way to map the cold gas component of a galaxy's structure.

Oxygen ($_8$O). Oxygen constitutes nearly two-thirds of all the atoms in Earth's crust. When bound to silicon it makes silicon dioxide (SiO_2), which is the active ingredient in sand, glass, and quartz. Oxygen is also the third most abundant element in the Sun and the universe. It is forged in the cores of massive red giant stars.

Contrary to popular belief, this element is *not* flammable. If you don't believe me, you can perform a simple test. Light a match. See if all the oxygen in Earth's atmosphere bursts into flames—of course this won't happen. However, oxygen promotes combustion while not being itself flammable. This is why the match burns and not the atmosphere. If Earth's atmosphere were 100 percent oxygen (it is only 21 percent oxygen), then when you lit the match the flame would spread rapidly to your body, and you would combust almost as fast as the matchtip. This is precisely what happened in January 1967 to Edward White, Roger Chaffee, and Virgil Grissom in the first Apollo space mission. While performing tests in their closed command module (filled with 100 percent oxygen) at their Cape Kennedy launch pad, a spark accidentally appeared. Everything burnable burned. The three astronauts were dead within seconds. The subsequent

Apollo missions were delayed so that the command module could be redesigned to contain combustion-resistant components and be closer to atmospheric percentages of oxygen.

That oxygen promotes combustion is not entirely unrelated to the utility of oxygen in the human body. Oxygen is the basis of animal metabolism—the ability to convert food into energy—which is not to be confused with animal magnetism.

Atmospheric oxygen is normally found with two atoms bound to each other, which is commonly written O_2. When a source of energy such as lightning is provided, atmospheric oxygen can add a third atom to become O_3. This form of oxygen is better known as ozone. The ozone layer resides in the upper atmosphere of Earth where it absorbs over 99 percent of the harmful ultraviolet rays, X-rays, and gamma rays from the Sun.

Neon ($_{10}$Ne). This inert element is most commonly seen in those skinny glass tubes that are used to advertise a product or service. The neon in these tubes glows orange. The entertainment and gambling mecca Las Vegas would be the city of Dark Shadows without it.

Sodium ($_{11}$Na). Symbol from the Latin *Natrium*. Like lithium (directly above it on the Periodic Table), sodium is a soft metal that will react violently with water.

Astronomers have met sodium in other ways. The biggest form of pollution to the mountaintop observational astronomer is not toxic waste or automobile fumes. It is street light. In the last quarter of the twentieth century, most American cities swapped the old-fashioned, dim, incandescent light bulbs for the bright, energy-efficient, high-pressure sodium vapor lamps. The sky is now brighter (more polluted) than ever before. One of the most heavily used observatory complexes in the world, the Kitt Peak National Observatory, is about sixty miles from Tucson, Arizona, and 7,000 feet above sea level. In an extraordinary response to this problem of light pollution, the entire city of Tucson agreed to use a special variety of low-pressure sodium street light that is far less polluting to sensitive astronomical measurements of stellar spectra. Only three thousand cities to go.

Silicon ($_{14}$Si). Silicon is a "kissing cousin" of carbon. Notice it immediately below carbon in the Periodic Table. The arrangement of silicon's outer electrons resembles those of carbon. For this reason, silicon and carbon are chemically similar. There is an entire branch of chemistry that investigates what happens when carbon atoms are replaced with silicon atoms in the panoply of molecules of the carbon repertoire.

One of silicon's many uses is that it is the ideal material in which to embed microcircuitry and electronics—hence the thought-provoking concept that modern computers are silicon-based life.

A minor revolution in astronomy began in the 1980s when ultra light-sensitive silicon-based detectors called charged coupled devices (CCDs) began to be used in place of older photographic detection methods. Astronomers now rely heavily upon CCD technology to detect the faintest of stars and galaxies out to the farthest reaches of the universe. During the Vietnam War, early versions of ultra light-sensitive detectors were used to assist military operations. Some American soldiers were equipped with specialized spotting scopes on nighttime "search and kill" missions. The searching part was aided by starlight.

Titanium ($_{22}$Ti). As strong as steel yet weighs only half as much, titanium is used in expensive bicycles and expensive airplanes. In astrophysics, most stars have more oxygen atoms than carbon atoms in their atmosphere. If the star also happens to be cool and red, then its visual spectrum is *dominated* by features that are attributed to the molecule titanium oxide (TiO_2). Note that the appearance of elements or molecules in a star's spectrum is a property of the star's composition *and* surface temperature. A hot blue star may also have titanium, but blue stars are too hot to make TiO_2, and titanium itself does not produce spectral features that are detected easily.

Iron ($_{26}$Fe). Symbol from the Latin *Ferrum*. Human blood chemistry is based on iron. Unbeknownst to the ancient Romans, blood is red for exactly the same reason that the planet Mars is red—the existence of iron-oxygen compounds. To associate the blood-red planet after their god of war was not only visually but chemically appropriate.

Even apart from its role in blood chemistry, iron may well be the most important element in the universe. The inward gravitational pressures of the highest-mass stars have no trouble raising their core temperature to fuse, in sequence, hydrogen to helium, helium to carbon, carbon to oxygen, and so forth up the Periodic Table. This cosmic transmutation from element to element is an alchemist's dream. At each stage, energy is released outward to support the gaseous star against further collapse. As a direct consequence, the star's core gets hotter and hotter, and at each stage a cooler shell of the previous nuclear fuel is left that surrounds the core. At the same time, for a variety of other complicated reasons, the outer layers of the star expand and cool so that the star becomes what is commonly called a red giant.

Eventually, the core fusion reactions arrive at iron surrounded by an "onion skin" shell structure of the cooler, previous fuel sources.

Up to now, the fusion of elements in the core has released energy and provided a sense of outward stability for the star. Stars are in the business of making energy. They know nothing else. But the energy that binds iron is *less* than the energy that binds any other element. If you fuse iron, the reaction will *absorb* energy. The star can no longer resist collapse, which occurs in just a matter of hours. All onion skin layers of the star spontaneously descend to an inner region. In a response not unlike a titanic bounce, these layers (formerly cool and safe) now undergo total and unconstrained thermonuclear fusion. The entire star blows its guts all over interstellar space in a supernova explosion. The luminosity of the star will increase a billion fold in a matter of hours. During the explosion, elements heavier than iron are formed. From this enrichment comes planets, asteroids, exotic molecules, and life itself. Yes, we are all stardust from some distant cosmic catastrophe. Blame it on iron.

Nickel ($_{28}$Ni). Ordinary nickel is the stuff that one might suppose to be the contents of the U.S. five-cent piece. But there is a more astrophysically interesting version of nickel that contains two fewer neutrons than everyday nickel. This neutron-deficient nickel is unstable. During a supernova explosion, there is a phase where the luminosity of the supernova is driven entirely by the energy that is liberated when nickel decays to cobalt.

Nickel is magnetic, as are iron (Fe) and cobalt (Co), to its left in the Periodic Table. Note that the U.S. coin known as the "nickel" does not even quiver when a household magnet is brought near it. These elements, when alloyed with aluminum (Al), make what are called "alnico" magnets, which are among the strongest in the world.

Copper ($_{29}$Cu). Symbol from the Latin *Cuprum.* In the *Star Trek* television and movie series, the blood of Mr. Spock (the pointy-eared, emotionless halfbreed Vulcan) is based on copper. Since copper turns green when mixed with oxygen, Mr. Spock has green blood. Copper is also cheap, and it conducts electricity very well, so electrical wires are mostly copper.

Zinc ($_{30}$Zn). One of those "minerals" that are supposed to be part of a healthy diet for humans, its presence aids in the digestion of proteins and the removal of carbon dioxide (CO_2) from the blood. It is also a major ingredient in U.S. coins, pennies included.

Arsenic ($_{33}$As). One of those "minerals" that could kill you posthaste, this used to be an active ingredient in murder mysteries.

Bromine ($_{35}$Br). Bromine is of no special astrophysical interest. Linguists, however, may appreciate that the name is from the Greek *bromos,* which translates to "evil-smelling."

Krypton ($_{36}$Kr). An inert, stable gas that, of course, resists combining with all other elements. Don't tell anyone, but there is no chemical reason why krypton-*ite*, even if it really existed, should trigger bad reactions in Superman.

Strontium ($_{38}$Sr). There is a radioactive version of strontium (the notorious strontium 90, with 38 protons and 52 electrons) that is one of the "waste" products of the atmospheric atomic bomb tests of the late 1940s and the 1950s. When farm animals grazed the contaminated grass in the vicinity of these test sites, their bodies absorbed the strontium in place of calcium. Notice that these two elements appear above and below each other in the Periodic Table; they are chemically similar. Hence, all the calcium-rich food products derived from these farm animals (milk, cheese, bone meal, and so on) were contaminated by radioactive strontium. Humans aren't farm animals, but chemistry is chemistry. The human body can't tell the difference between calcium and strontium either.

Yttrium ($_{39}$Y). One of those elements that looks hard to pronounce, Yttrium, along with Terbium ($_{65}$Tb), Erbium ($_{68}$Er), and Ytterbium ($_{70}$Yb) are elements named for Ytterby, a sleepy village near Stockholm, Sweden. The town has a quarry of unusual minerals that holds deposits of these previously unidentified elements.

Molybdenum ($_{42}$Mo). When this element is combined with manganese, you get the tongue-twister manganese-molybdenum. This is a very light, rigid steel alloy that is a favorite of bicycle frame builders.

Technetium ($_{43}$Tc). If you look up the prefix "tech-" in the dictionary, you will see synonyms such as "artificial" or "man-made." Technetium is radioactive and is not found naturally anywhere on Earth. It was the first element to be artificially produced. In spite of its nonexistence on Earth, technetium has appeared in the spectra of certain cool red stars. But technetium is unstable—if you have time to watch a pile of it, you will notice that half of it will decay within 200,000 years. We readily conclude that technetium was not present when the star was born typically millions or billions of years ago. It must have recently come from the interstellar medium or, more likely, it must have formed near the star's surface. To use unstable elements for this kind of deduction forms a very powerful tool to constrain theoretical models of stellar composition and evolution.

Silver ($_{47}$Ag). Symbol from the Latin *Argentum*. Silver is an even better conductor of electricity than copper. To keep most people from digging up the streets to steal power lines, electric companies wisely use copper instead.

Tin ($_{50}$Sn). Symbol from the Latin *Stannum*. "Tin" cans are actually made of steel (with a micro-thin layer of tin on the inside). Household "tin" foil is, of course, made of aluminum.

Iodine ($_{53}$I). Yes, this *is* the active ingredient in the antiseptic reddish stuff that your parents may have put on your bruises and made your injury look twenty times worse. But advances in medicine are such that if you were born after 1965 your childhood injuries were probably treated with topical antibiotics rather than iodine.

Tungsten ($_{74}$W). Symbol from the German *Wolfram*. Tungsten has the highest melting point of any element: 6,170° Fahrenheit. Thus, there is no better choice for the filament of an ordinary light bulb. It is also used in the contact points of a car's distributor cap and in armor-piercing bullets.

Osmium ($_{76}$Os). Here is the densest element on the entire Periodic Table. A cube of it that measures one foot on each side weighs about two thousand pounds. Osmium would, of course, make an ideal gift as a paperweight.

Iridium ($_{77}$Ir). Another dense element, a cubic foot of iridium would weigh only a few pounds less than a cubic foot of osmium.

Platinum ($_{78}$Pt). Osmium, iridium, and platinum are the three densest elements. They fall consecutively in the Periodic Table, as do iron, cobalt, and nickel, the magnetic threesome two rows above.

The international standard meter used to be the length of a 90 percent platinum–10 percent iridium alloy bar held at constant temperature in the International Bureau of Weights and Measures in Sèvre, France. In 1960, however, the meter was redefined (so that you cannot look at it or touch it) by international agreement to be 1,650,763.73 wavelengths of an orange-red spectral feature in the element krypton–86. With this definition, the length of the meter could be reproduced in any appropriately equipped laboratory. But now that the speed of light has been set, the *exact* length of the meter (as noted in chapter 3) becomes the distance light travels in 1/299792458th second—although this definition is somewhat less practical for use in the laboratory.

Gold ($_{79}$Au). Symbol from the Latin *Aurum*. The last I checked it was several hundreds of dollars per ounce. Gold is remarkably resistant to corrosion. It was the metal of choice for the engraved information plaque that was attached to the interplanetary space probe *Pioneer 10*. The gold plaque uses symbols to convey among other things the point of origin of the space probe, the number of planets in the solar system, and the relative size of humans to the probe's antenna. Incidentally, *Pioneer 10*, launched in March 1973, has now passed the orbit of Pluto and is on its way to the depths of interstellar space.

Gold is also quite dense. If you managed to toss a standard Fort Knox–style gold brick at your best friend's chest, you would probably break some ribs.

Mercury ($_{80}$Hg). Symbol from the Greek *Hydrargyrum*. Mercury is a liquid at room temperature. It is quite fun to play with because it balls up and rolls around the floor like little liquid BB's from a BB gun. But mercury can be dangerous. If it enters your blood stream it can induce, among other things, severe personality changes. Senility is inevitable in advanced stages of mercury poisoning. Sir Isaac Newton liked to play with mercury. (Actually, he spent a big part of his life playing with chemical elements.) Some historians have hypothesized that mercury poisoning led to Sir Isaac's reduced scientific output in his later years. We should be grateful that he did not play with mercury until *after* he invented calculus, discovered the laws of gravity and optics, and invented the reflecting telescope. This should be a lesson to all of us.

Lead ($_{82}$Pb). Symbol from the Latin *Plumbum*. Lead appears in the chapter on Venus in many introductory astronomy textbooks. At 900° Fahrenheit, the surface of the beautiful planet Venus is well above the temperature required to melt lead. Lead is also poisonous which is why it is no longer found in household paint and why it is no longer used in America as an octane-boosting gasoline additive. Lead poisoning in a pregnant woman can lead to birth defects, and if lead is ingested by a growing child then physical and intellectual growth may be stunted.

Radon ($_{86}$Rn). Like other radioactive elements, this inert, unstable, radioactive gas wants to decay and release high-energy gamma rays. These are the kind of rays that can induce leukemia and mutate your genes. Radon emerges naturally from soil and is therefore common in the environment. It can thus pose health hazards for people who live in super-insulated homes, where radon gets trapped as it emanates from the earth beneath the house.

Uranium ($_{92}$U). Named for the planet Uranus, the seventh planet from the Sun. An unstable isotope of uranium with 143 neutrons was used by the United States in 1945 to kill over 50,000 Japanese residents of Hiroshima in the first atomic fission bomb ever dropped in warfare. Another 50,000 died later ⸗from delayed physiological reactions to high-energy radiation exposure. If you collide free neutrons with uranium, the nucleus will break apart into several lighter elements and liberate more free neutrons that serve to break apart other uranium atoms. In a specially designed enclosure, a chain reaction ensues, which releases an enormous amount of energy. This is called "fission", which is the basis of all twentieth-century nuclear power plants.

The energy liberated in a fission reaction was originally postulated as a possible energy source for the Sun. But the rarity of uranium and other large elements, and the availability of hydrogen and helium, has led to the more realistic and correct view that thermonuclear hydrogen fusion powers the Sun rather than fission.

Neptunium ($_{93}$Np). Named for the planet Neptune, the eighth planet from the Sun. All versions of this element are unstable, but it can be found in trace amounts in uranium ores.

Plutonium ($_{94}$Pu). Named for the planet Pluto, the ninth planet from the Sun. Plutonium, like neptunium, can be found in trace amounts in uranium ores. Plutonium may also be considered the largest naturally occurring element.

Apparently, once was not enough. The second and last atomic bomb ever used in warfare was in 1945—three days after the Hiroshima bomb—when the United States dropped a plutonium fission bomb on Nagasaki, Japan, killing about 20,000 residents. This death toll, as in Hiroshima, nearly doubled after deaths from radiation exposure were tabulated. Many descendants from the survivors of Hiroshima and Nagasaki were born mutated and disfigured. The casualties of war extended to the unborn.

Americium ($_{95}$Am). If there had been a known tenth planet, this element might have been named for it.

Mendelevium ($_{101}$Md). It took 101 elements but Mendeleev, the "father" of the Periodic Table, was finally honored with this unstable element.

Unnilquadium ($_{104}$Unq), Unnilpentium ($_{105}$Unp), Unnilhexium ($_{106}$Unh), Unnilseptium ($_{107}$Uns), Unniloctium ($_{108}$Uno) and Unnilnonium ($_{109}$Unn). None of these heaviest and most recently isolated elements "live" for longer than about a minute (give or take a moment or two). They are created out of atomic curiosity in nuclear laboratories. I give their names primarily because they sound funny. After some suggestions are put forth by their discoverers, these elements will ultimately be renamed for a person, place, or thing by the International Union of Pure and Applied Chemistry.

The truly great ideas of science tend to be both aesthetically beautiful and profound. It is no surprise that these are attributes shared by the great works of poetry. If you are a poet and you have read this far, then perhaps you will confess that the Periodic Table of the Elements is no less inspired than your periodic lines of words.

PART THREE

·

*Astronomy Is
Looking Up*

· *11* ·

Menagerie

There are eighty-eight keys in a piano, and there are eighty-eight constellations in the sky. The eighty-eight piano keys make music. The eighty-eight constellations make a zoo composed of one insect, one crustacean, one angry arachnid, five fishes (with a pair among them), five reptiles, nine birds, three women, twelve men (with a set of twins among them), five canines (inclusive of a hunting duo), fourteen other mammals, five mythical-magical creatures, and thirty inanimate objects that include three boat parts, ten scientific instruments, one musical instrument, two crowns, a flat-topped mountain, somebody's hair, and a river.

‿

To supplement your nighttime viewing, here is some underpublicized information that a well-informed stargazer should know.

From a species point of view, the following constellations are in the record-book of celestial creatures:

Tallest:	Camelopardalis, the Giraffe.
Heaviest:	Hydra, the Whale
Smallest/Lightest:	Musca, the Fly
Most Poisonous:	Scorpius, the Scorpion
Fastest:	Pegasus, the Winged Horse
Strongest:	Hercules
Prettiest:	Pavo, the Peacock
Ugliest:	Medusa's snake-ensnarled bloody severed head, as displayed by Perseus

From a connect-the-dots point of view, the constellation Orion has the rare combination of large size, bright stars, and an outline that resembles the hunter he is purported to be. His neck, shoulders, waist (belt), knees, sword, and shield are all clearly defined. Unfortunately, he hasn't much of a head—there is a big empty space above his neck. There is some controversy about whether Orion is left-handed or right-handed. Early drawings and woodcuts from the fifteenth, sixteenth, and seventeenth centuries show the back of Orion's head, his rear end, and the rest of his loin-cloth-draped body facing away. The star pattern requires that he wield his wooden battle-club with his left hand, which makes Orion the world's largest and most famous lefty. Illustrated globes of the celestial sphere from the same period (an excellent collection may be found at the Musée

National des Techniques in Paris) also depict Orion from the rear, even though the constellations are intended to be viewed from the "other side" of the sky and thus should be drawn in reverse. More recent sketches of Orion (probably drawn by righties) show him face on as he wields his club in his right hand.

Orion's sword is commonly illustrated over a short string of stars that hangs from his belt and dangles between his legs. I have never hunted with a sword and club, but of all the places on my anatomy that I might carry a sword, it seems to me that between the legs would be low on my list. Such is the cost of connecting the dots.

The stars in Pegasus, the flying horse, are not quite as bright as those in Orion, but they are just as majestic. Clearly visible are four stars of the "Great Square" that form the horse's body. Front legs drape below it. Extending forward is a slightly bent line of stars that resembles the curve of a horse's neck and head. You must rely on your imagination for its wings. It is not commonly discussed that Pegasus is only a half a horse. You must also invoke your imagination if you wish to picture Pegasus's rear end, because the constellation Andromeda occupies the region that would otherwise complete the horse. By coincidence of configuration, the interior of the Great Square of Pegasus is remarkably devoid of visible stars—the square is as impressive for its near-square geometry as it is for its emptiness. And unbeknownst to our empty-bellied winged steed, Pegasus flies through the sky upside down as viewed by residents of the Northern Hemisphere. I am convinced that if constellations had been first identified by Americans of the late twentieth century rather than by ancient civilizations, the the Great Square of Pegasus would be known as the the "Great Television" in the sky.

The award for the most exotic star names must go to the

otherwise undistinguished constellation Libra, the Scales. Its two brightest stars are officially named Zubenelgenubi and Zubeneschamali.

The most boring constellation in the sky is no doubt Triangulum Australis, the Southern Triangle. A detailed photograph of its three brightest stars shows—you guessed it—a triangle. Since nearly any three stars in the sky form a triangle, Triangulum gets the award for the most unimaginative constellation name. To be fair to Triangulum, there are several dimmer stars in and around the triangle. But since the constellation is simply the "Southern Triangle," these stars do not participate in the designated pattern.

The greatest stretch of the imagination occurs with Apus, in the Southern Hemisphere. It is a constellation with three prominent stars near the South Celestial Pole that is supposed to be a fully plumed bird-of-paradise.

Some stars grow in the mind. The most famous of these is Polaris, the North Star. In an informal poll, I once asked passers-by, "What is the brightest star in the nighttime sky?" Three-fourths of them unwittingly proclaimed, "The North Star!" Let it be known that the North Star is not even in the celestial top forty. In addition, its reputation puts it at the point in the sky that is directly over the Earth's North Pole. In the real sky, however, Polaris is nearly one degree from the North Celestial Pole—about twice the width of the full moon. I do not wish to upset anybody, but in 12,000 years, due to the wobbling of Earth's axis, Polaris will be over forty-five degrees from the celestial pole. Perhaps our North Star should be renamed Somewhere-Near-the-North-Pole-Aris. In spite of all this, residents of the Northern Hemisphere should not complain. Currently, the region of sky that surrounds the South Celestial Pole is practically blank. The nearest star with a brightness similar to that of Polaris is over twelve degrees away.

For the record, the brightest star of the nighttime sky is Sirius (Alpha Canis Majoris) in Canis Major, the Big Dog. It is nearly thirty times brighter than the North Star and commonly depicted as the gleam in the Big Dog's eyeball. Indeed, Sirius is affectionately known as the "Dog Star." Sirius is quite recognizable as it lurks below and to the left of Orion. Sirius is also visible from nearly the entire inhabited Earth during one season or another, but it is best viewed in December and January when it rises at sunset and sets at sunrise. A Sirius joke between stargazing astronomers occurs when, after hearing an unbelievable story, one declares to the other, "You can't be serious!" The response: "No, I am not Sirius, I am Zubenelgenubi!" At the end of July, Sirius rises just before the morning Sun, as though the Sun were walking its dog into the summer sky. This annual celestial ritual thus heralds the onset of the hot-and-steamy "dog days" of August.

The appearance of Sirius just before sunrise was historically well-timed with the annual rise of the Nile River through Egypt and thus became a harbinger of a renewed agricultural cycle. So important was (and is) the rising Nile to life in Egypt that the 5,000-year-old Egyptian calendar uses the appearance of Sirius just before sunrise as the first day of the year.

Sirius is actually a double star system. The dimmer of the pair, now called "Sirius B," is an extremely dim degenerate white dwarf. Its existence was not telescopically confirmed until 1862, when Alvan G. Clarke, an ace observational astronomer, revealed its presence buried within the glare of "Sirius A."

The nearest star to Earth, as conclusively established by extensive astronomical research, is the Sun. It is often declared that the nearest star to the *Sun* is Alpha Centauri, the brightest star in the southern constellation Centaurus and the third brightest star in the night sky. Alpha Centauri is, however, a double star system, and neither star in the pair is the closest star to the Sun. That privilege goes to the dim star Proxima Centauri, which is near enough to the Alpha Centauri pair to complete an orbiting triple star system. All three stars compose the front hoof of the Centaur as he straddles the Southern Cross. At one hundred times dimmer than the detection limit of the naked eye, Proxima Centuri makes a rather demure nearest neighbor.

The constellation with the greatest hype is Crux Australis, the Southern Cross. There are songs written about it, and it appears on the national flags of Australia, New Zealand, Western Samoa, and Papua New Guinea. What they do not tell you is that the constellation is small—it is the smallest of all the eighty-eight. Indeed, your fist at arms length would eclipse it entirely. Its four brightest stars outline the corners of a crooked square, or a kite. In geometric terms it is nearly a rhombus, although "Southern Cross" does convey more romance than "Southern Rhombus." There is not even a star in its middle that could represent the center of a cross. The Southern Cross is best used as a signpost to find other, more interesting celestial objects. For example, the Southern Cross is thirty degrees north of the star-starved South Celestial Pole, and ten degrees southwest of the titanic naked-eye globular cluster Omega Centauri. And the Galactic equator, also known as the Milky Way, passes directly through its middle.

Two relatively recent additions to the celestial menagerie are the southern constellations Telescopium and Microscopium, the Telescope and the Microscope. Unlike Triangulum Australis, which is simply boring, each of these two constellations are boring *and* undistinguished. The brightest stars in Telescopium and Microscopium are over one hundred times dimmer than Sirius. These constellations date not from the ancients but from Abbé Nicolas Louis de la Caille of the middle eighteenth century. With decidedly less imagination than the ancients, La Caille identified fourteen new groups of stars from the poorly charted southern celestial sphere. He honorably named them for the principal instruments (hardware) of the arts and sciences. As noble as all this sounds, La Caille had no excuse, and thus is never to be forgiven, for naming two of the least distinguished constellations in the heavens after two of the most important scientific instruments of our times.

A constellation that was simply too big for its neighborhood was the sprawling Southern Hemisphere constellation Argo Navis, or Argo the Ship. Its length spanned nearly one-fifth of the entire sky. Mythology holds that this is the same ship made famous by Jason and his fifty Argonauts, who set sail from Iolchis in Thessaly to Aea in Colchis to search for the golden fleece. The disproportionate size of Argo Navis led our friend Abbé Nicolas Louis de La Caille to cut up the constellation into four smaller patterns while preserving the boat theme. Thus was born Carina the Keel, Puppis the Stern, Pyxis the Compass, and Vela the Sail.

Enduring favorites for the three quarters of the world's population who live in Earth's Northern Hemisphere are the Big Dipper and the Little Dipper. They are officially "asterisms," which simply means that they are interesting subsets of otherwise uninteresting constellations. The Big Dipper's seven stars form a convincing kitchen saucepan in the sky: three stars form the slightly curved handle, four stars form the pot. Incidentally, the two stars of the saucepan's front edge are reputed to point toward Polaris, but they miss their target by nearly three degrees. Hanging off Polaris is the Little Dipper. Its handle is curved the other way when compared with the Big Dipper. It looks very much like a cauldron ladle with Polaris at the handle's tip.

The Big and Little Dippers are actually parts of the constellations Ursa Major and Ursa Minor, the Big and Little Bear. They are reported to be rather chubby bears (as bears are wont to be) with long bushy tails that also form the handles of the saucepan and ladle. But these long tails are actually part of cosmic tales because tails of terrestrial bears are only nubby stubs.

Keeping with the kitchen theme, we go to an asterism in the constellation Sagittarius. Sagittarius is a centaur-archer who is part man and part horse (the *front* end is the half man). In spite of this legendary description, the brightest stars bear a remarkable resemblance to a stove-top tea kettle. It is short and stout—complete with a handle and a spout. This asterism is especially revered in England because the band of light from our Milky Way galaxy appears to pass through the tea kettle's spout. In England, they always take a spot of milk in their tea. In China, however, milk was never a popular beverage. The Chinese know the Milky Way as "Yin-hur," or Silver River. Aside from its kitchen-accessory status, Sagittarius is deservedly famous because it contains the center of the Milky Way galaxy—located about three degrees west of the spout.

The most misidentified asterism in the sky is the Pleiades. This little bunch of seven stars has a vague resemblance to a dipper. Since it is little (your thumb held at arm's distance would cover all visible stars), many people mistakenly call it the Little Dipper. The Pleiades may be found above and to the right of Orion's missing head. In Greek legend the seven stars of the Pleiades represent the seven daughters of Atlas: Alcyone, Maja, Merope, Taygete, Asterope, Electra, and Celeno. While a simple telescope shows dozens of stars, the naked eye sees only six. Celeno is missing. To reconcile this numerical error the fourth-century Alexandrian-Greek commentator Theon the Younger surmised that Celeno, which is the dimmest of the group, must have been struck by lightning.

To experienced stargazers, the constellation with the most convincing resemblance to a letter of the alphabet is Cassiopeia, queen of Ethiopia. She owes her celestial existence to five bright stars in the sky that form a W, which according to some legends is her throne. The W is somewhat lopsided, like a chair that is ready to collapse—so it is rumored that she gained weight in her later years. Cassiopeia is near enough to the "pole" star Polaris that for most of the Northern Hemisphere she never sets. At various times of the night and at various times of the year, she can be found above, below, and to each side of Polaris. The W will sometimes be an Σ (the upper case Greek letter sigma), sometimes an M, and sometimes a $\exists$. This merry-go-round behavior is not a fitting fate for a queen, but Cassiopeia once said she was more beautiful than the Nereids (the Water Nymphs). The gods did not take kindly to this boasting and (among other things) condemned her to swing eternally around the pole.

As detailed in the next chapter, the constellations with the greatest irrational following are the twelve of the zodiac: Aries, Taurus, Gemini, Cancer, Leo, Virgo, Libra, Scorpio, Sagittarius, Capricorn, Aquarius, and Pisces. One is often led to believe that the zodiacal constellations are prominent in the nighttime sky. But astrologers do not tell you that Aries, Cancer, Virgo, Libra, Capricorn, Aquarius, and Pisces are underwhelming constellations that are barely recognizable as coherent patterns in the nighttime sky. Astrologers also do not tell you that the constellations are not the same size, so that the Sun does not move across them at equal one-month intervals. Astrologers further fail to tell you that the correspondence of the zodiac with calendar months is shifted backward by an entire constellation due to Earth's ongoing precession on its axis. And finally, astrologers do not tell you how much money they make from gullible people.

The fact remains: all you ever see in a clear night sky is a few thousand dots of light. If you would like to see a real menagerie, and you cannot hallucinate like the ancients, then visit your nearest zoo. You will see real (tailless) bears, real (wingless) horses, real scorpions—but, alas, no centaurs. These animals will look exactly as nature intended. And the zookeeper will not give you advice about your financial life, your home life, or your love life.

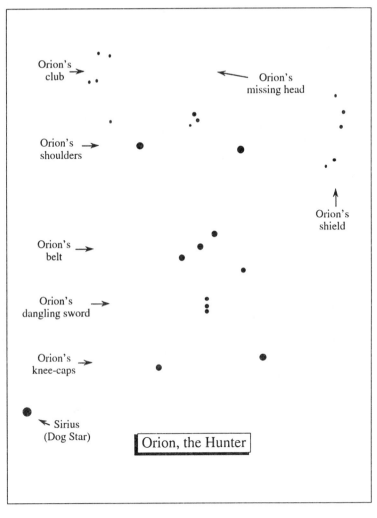

Orion's club

Orion's missing head

Orion's shoulders

Orion's shield

Orion's belt

Orion's dangling sword

Orion's knee-caps

Sirius (Dog Star)

Orion, the Hunter

Figure 11.1

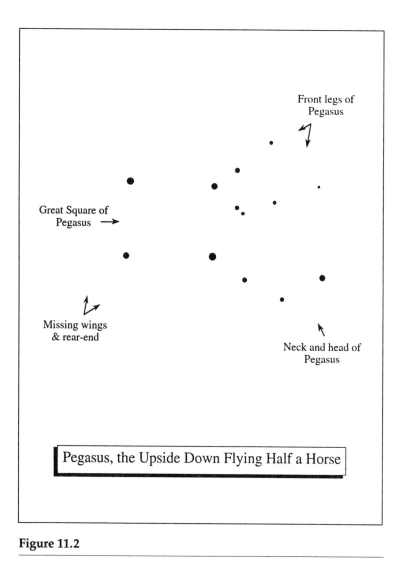

Figure 11.2

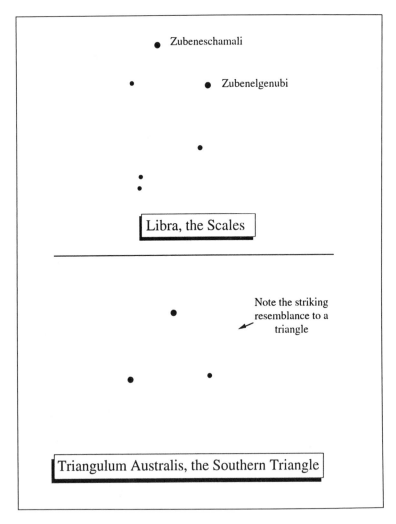

Figure 11.3

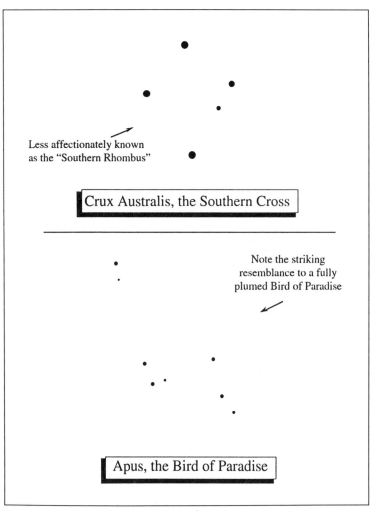

Figure 11.4

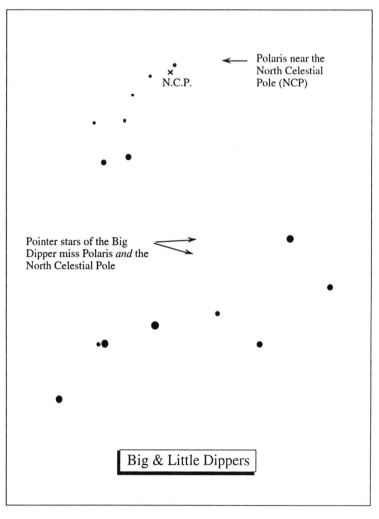

Figure 11.5

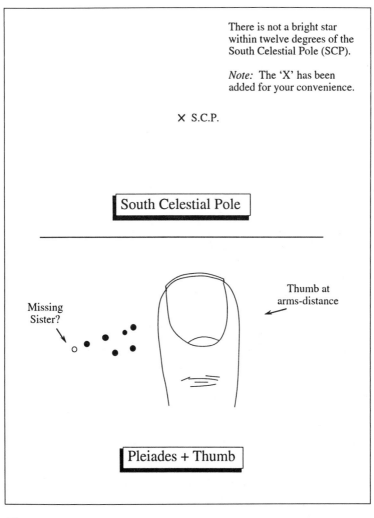

There is not a bright star
within twelve degrees of the
South Celestial Pole (SCP).

Note: The 'X' has been
added for your convenience.

✕ S.C.P.

South Celestial Pole

Missing
Sister?

Thumb at
arms-distance

Pleiades + Thumb

Figure 11.6

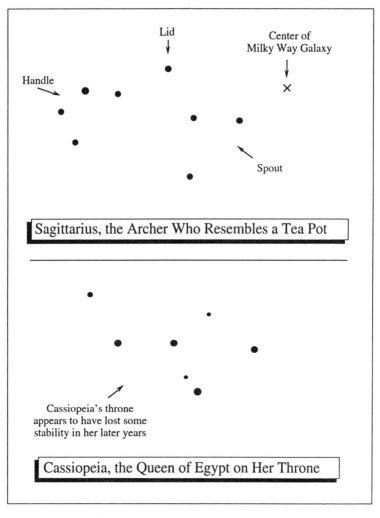

Sagittarius, the Archer Who Resembles a Tea Pot

Cassiopeia's throne appears to have lost some stability in her later years

Cassiopeia, the Queen of Egypt on Her Throne

Figure 11.7

· *12* ·

Horrorscope

The Sun is among hundreds of billions of stars in the disk of our own Milky Way spiral galaxy. The stars in the Sun's vicinity comprise what is politely called the solar neighborhood. Some stars emit copious amounts of light while others emit much less. Some are near and some are far. Yet all stars present the illusion that they are firmly embedded in the "dome" of the night sky. Any two stars form a line. Any three stars form a triangle. Imagine the endless shapes and patterns that can be envisioned among the six thousand stars visible to the unaided eye.

⌒⊙⌒

*M*ore than five millennia ago, in the days before evening television, the Babylonians and Chaldeans kept accurate records of the night sky. They assigned icons of the local mythology to the planets and the various star patterns. Between A.D. 127 and A.D. 150, the Greek philosopher Ptolemy catalogued many of the familiar names upon these skeletal star patterns that are otherwise known as constellations. Thrust upon these patterns were not only the names of animals and gods but their quirky behavioral traits as well.

More than two millennia after Ptolemy, in the days of space travel, recombinant DNA, and microwave popcorn, there are people who believe that planets and star patterns influence their life's events in ways derivable from the mythology of sleepless and TV-less Babylonians.

The basis for modern-day astrology is quite simple. Its premise is that the relative positions of the Sun, Moon, planets, and constellations affect you and the events in your life—especially your social and financial life. Note that in a free, capitalistic society these aspects can be the totality of one's life. All planets of the solar system orbit the Sun in roughly the same plane. Consequently, it makes sense that when viewed from Earth, the motion of the planets, Moon, and Sun appear to be restricted to a relatively narrow band across the entire sky. The exact path of the Sun is centered in this band and is known as the "ecliptic." As Earth orbits the Sun, we see the Sun travel eastward along this ecliptic until one year has passed when the Sun returns to its initial position. If Earth's rotational axis were not tilted, the ecliptic would coincide with the projection of Earth's equator on the celestial sphere. Our 23½-degree tilt, however, creates what is tongue-twistingly called the "obliquity of the ecliptic." This obliquity grants two special points to the Sun's yearly journey across the sky—both occur where the ecliptic crosses the projection of Earth's equator. They are commonly called the first day of spring (the vernal equinox) and the first day of autumn (the autumnal equinox). The first day of spring is referred to by astrologers as the "first point of Aries," which officially begins the Sun's journey through the twelve constellations of the zodiac. If your astrological sign is Aries, this means the Sun was passing through the constellation Aries when you were born. Following Aries on the calendar are Taurus, Gemini, Cancer, Leo, Virgo, Libra, Scorpius, Sagittarius, Capricornus, Aquarius, and Pisces.

Many astrologers say they can predict your behavior and personality based on your sign in conjunction with the signs that the planets and Moon happen to be in when you were born. It would be wrong to say that the planetary positions in the sky have *no* effect on human behavior. Planets do reflect light from the Sun and do have gravitational forces. Both factors (the only quantities measurable from Earth) have the potential to affect objects at a distance. You should know, however, that if we calculate the force of gravity of, say, Mars on you when you were born, you will discover that the force of gravity from the obstetrician or midwife who delivered you was *150 times* greater than that of Mars.

This leaves light as a last hope for the effects of Mars on your birth. If we presume optimum conditions: you were born on a clear night by an open window in a hospital that happened to have Mars in the sky and in view from the delivery table, then we may discuss the effects of its light. We discover here that the six high intensity lamps over your delivery table produced *160 billion times* more light on you than the light from Mars. It is not clear why national cults have never emerged that would predict your fate based on how the obstetrician was positioned at your birth, or whether GE bulbs were used instead of Westinghouse bulbs.

Astrologers generally agree that the most important astrological effect is the Sun's location on the zodiac at your time of birth inasmuch as it defines your "sign." But those who believe in the Sun's positional influences will have to contend with a sobering fact: the first point of Aries today no longer coincides with the first day of spring. This is a far-reaching twist of the same magnitude as if Alex Haley, author of his well-known genealogy *Roots*, had later discovered that he was an adopted child. The Sun at the vernal equinox is currently in the constellation Pisces and will soon be in Aquarius. Among people who read their daily horoscope, it is not widely known that Earth wobbles on its axis. This is normal behavior for any oblate, spinning, tilted top under the influence of an external gravity, so we should not be surprised that the spinning tilted Earth also wobbles on its axis under the influence of the Sun and Moon. The time for one complete Earth wobble is about 25,700 years, and its effects are manifested by the drift of the vernal equinox through the entire zodiac over a 25,700-year period. Ptolemy named the constellations about two thousand years ago. We see that, since then, the equinox has traveled nearly one-twelfth the way around the zodiac—or one complete sign. What this means is that if you thought you were an Aries, you are *really* a Pisces; if you thought you were a Pisces, you are *really* an Aquarius, and so forth. Matters are worsened upon learning that the boundaries of the constellations do not split the zodiac evenly—some astrological signs "last" longer than others. Matters are worsened further upon recognizing that seven of the twelve constellations are feeble skeletal excuses for the animals and objects they are purported to be.

An amusing addition to the above high jinks is that the zodiac contains fourteen constellations, not twelve. The Sun, after leaving the constellation Scorpius, enters the constellation Ophiuchus. It then stays in Ophiuchus for a *longer* period of time than Scorpius, the sign that is advertised to precede Sagittarius. The confusing conclusion is that most Scorpions are actually Ophiuchans, and all Scorpions and Ophiuchans are currently Librans. The fourteenth constellation in the set is Cetus. It is a large constellation that dips into Pisces. The Sun passes through Cetus briefly as it ambles through Pisces, but you are not normally informed of this in the horoscope pages.

Some astrologers (typically the expensive ones) are actually aware of these astronomical truths. A common response, when confronted with the facts, is the assertion that the effects of the stars were set two thousand years ago and still apply today. I once tested a daily (syndicated) horoscope from a local newspaper on the fifty students in one of my introductory astronomy classes. Rather than have the students read their own horoscope and decide whether it applied to that day's dilemmas, I picked one of the twelve horoscopes at random and read it to the class. I then asked all students to declare whether it was "unlikely," "possible," or "likely" that I had just read their own horoscope. Fully one third (seventeen) of the class declared that the horoscope was "likely" to be theirs. The class was astonished to learn that the horoscope I read belonged to *none* of these people. Of the ten people who responded "unlikely," the horoscope actually belonged to three of them. Controlled experiments such as this one consistently demonstrate that daily horoscopes would do no worse if they were laid on the page at random, yet horoscope casting in the United States remains the most lucrative industry among the pseudosciences.

. . .

Welcome to Beebe Library!

Where Wakefield Connects

wakefieldlibrary.org | (781) 246-6334

You checked out the following items:

Patron Name: Doherty, Theresa
Patron barcode: 21392000837241

1. **Universe down to Earth**
   Author: Tyson, Neil De Grasse.
   Barcode: 31607001453740
   Due Date: 1/13/17 11:59 PM

By borrowing these items, you saved
$0.00!

...nted on 12/30/16 11:00

A subject of fascination and confusion for many people is the effect of the full Moon on human behavior. It is commonly thought that more babies are born during full moons than during any other phase. It is also thought by many people that the full moon has some mystical effect on the human psyche that forces people to behave strangely, to commit crimes, or to transmutate into a howling and hairy canine. Literature abounds with stories of werewolves and other moon-induced human behavior. I have even heard some people explain these phenomena with the tidal effects of the Moon on the brain: "Since the oceans are water and are duly influenced by the lunar tides, then the large water content of the human body should also be affected."

Before we jump to cosmic conclusions, consider the following: (1) If the oceans were 100 percent nitroglycerin (or 100 percent anything else), they would still exhibit tides. (2) Tidal forces of the Moon are indeed large during full moon. But they are also large two weeks later during the new moon. This phase of the moon cannot be observed. Nobody sees it. Nobody writes lycanthropic stories about it. (3) Tidal forces of the Moon are measured by the difference in gravity between the side of Earth closest to the Moon and the side of Earth farthest from the Moon. If your skull were 7,000 miles across (the size of Earth), then the lunar tides would indeed give you an oblong-shaped head, with untold consequences on your mental facilities. But since your skull is only about eight inches across, the tidal force that you "feel" is quite small. Indeed, the weight of a down-filled bed pillow placed upon your head will produce a force that is seven trillion times larger than that of the Moon's tidal force on your head. So the next time somebody tries to blame a bio-cosmic-lunar connection for their irresponsible behavior, perhaps we should first blame the influence of creative literature—and then possibly the pillow.

Let's return to the subject of birth rates during the full moon. The average human gestation period is about 267 days,[1] which happens to be an excellent match with the 265½ days in nine cycles of lunar phases. What this means is that babies who are born during a full moon are very likely to have been *conceived* during a full moon—and nobody will argue the romantic effects of a moonlit evening.

In a free society, intellectual enlightenment is your best defense against misguided claims in the name of science. Only then can society, as a whole, cultivate a scientifically literate public.

1. You can get this from a commonly invoked yet convoluted formula: Add seven days to the first day of your last menstrual cycle before becoming pregnant. Count backward three months, and then add a year. If we assume a 28-day menstrual cycle, and conception during ovulation, then this formula will give you 267 days.

· 13 ·

Celestial Windings

I know that I am mortal by nature, and ephemeral; but when I trace, at my pleasure, the windings to and fro of the heavenly bodies I no longer touch earth with my feet: I stand in the presence of Zeus, himself, and take my fill of ambrosia.

—Claudius Ptolemy

*U*niversity astronomy departments and planetariums, especially those near large population centers, typically receive hundreds, sometimes thousands, of daytime telephone calls per year from the general public with questions about cosmic phenomena. Some of the calls are induced by heavily publicized events such as lunar and solar eclipses, or planet-moon conjunctions, while other telephone calls are simply the consequence of people with curious minds who should have otherwise been busy at their jobs. In all cases, however, the array of questions reveals a genuine interest in celestial happenings that serves as a daily reminder to professional astronomers that in the absence of telescopes and computers and theories, one can still be awed by just looking up.

Earth

It is often said that Earth's axis is tipped in space. But in space there is no uniform up or down, so being tipped can only have relative meaning. We can draw on a sheet of paper the slightly flattened circle of Earth's eccentric orbit and ask whether Earth's axis points straight out of the page. It does not. Earth's axis is tipped slightly more than one-fourth of the way toward the plane of the page. When measured in angle, it amounts to about 23½ degrees. That the round Earth rotates on a tipped axis and revolves around the Sun required millennia of the world's greatest thinkers to unravel. So there is no need to get upset if this circus of motion has ever left you confused.

It is sometimes convenient to think of the sky above you

as the inner surface of an inverted salad bowl, which forms what is otherwise known as a hemisphere. Following this analogy, the entire sky as seen from Earth is known as the celestial sphere. By helpful coincidence, the North Pole of Earth's axis points near a star "on" the sky, which is, of course, called the North Star. The South Pole points to a big empty area that is not too far from the Southern Cross. If we continue this cosmic correspondence, we can also project Earth's equator onto the sky. With this simple exercise, we have identified three places: the North *Celestial* Pole, the South *Celestial* Pole, and the *Celestial* Equator. In a layout that is analogous to Earth's longitude and latitude, there exists coordinates for the sky called "right ascension" and "declination."

Contrary to popular belief, Earth rotates on its axis once in 23 hours and 56 minutes, not 24 hours. In other words, a star, or any other spot on the sky, will return to the same location above you every 23 hours, 56 minutes. On average, however, the Sun reaches its highest spot on the sky every 24 hours. For daily scheduling, people tend to respect, honor, and obey the Sun—not the rest of the stars in the sky. Most of human civilization has therefore chosen to set clocks against the 24 hours of the Sun. Astronomers, however, conduct business in star time. All timekeeping devices that are set to the stars are called sidereal clocks, where midnight sidereal time equals midnight Sun time only once a year—on the first day of autumn, which falls on or near September 21. Thereafter, for every day of the year, the sidereal clock will gain four minutes against the Sun clock because Earth must rotate an extra four minutes just to re-turn the Sun to the same location as the day before.

Earth's orbital motion ensures that day to day the Sun's position in the sky will migrate across the background of stars.[1] There is nothing complex about this. If your name were Fido, and you were tethered to a pole, and if you decided to run in circles around it, then you would systematically observe the pole to appear in front of every part of your surroundings. Earth is tethered to the Sun by gravity, and Earth moves in unending circles around the Sun. The only important difference is that Earth is not likely to strangle itself.

Longitude on Earth is measured in degrees, yet right ascension, the corresponding cosmic coordinate, is measured in hours. Where does right ascension begin? In the same place that longitude begins, at the Royal Greenwich Observatory in Greenwich, England. Using an accurate clock—sidereal, of course—the time in Greenwich *is* the right ascension of the star that happens to be crossing a line through the zenith that connects due north and due south. For anybody in the world, this line is called a meridian, but for Greenwich it is exaltedly known as the Prime Meridian—not by cosmic mandate but by international convention. Zero degrees longitude, the Earth boundary between east and west, is also defined to go through Greenwich. Incidentally, there is no cosmic reason why the Prime Meridian could not have been Eddie's Steak House in Kalamazoo, Michigan—except that Eddie would be obligated to supply right ascensions to the world astronomical community for all stars in the sky. He could, however, start a catchy ad campaign, "Enjoy your Prime Rib on the Prime Meridian!"

Sometimes simple longitudes, latitudes, and meridians are not enough. I once received a telephone call at my office from a practicing Muslim who was new to the New York City area. The caller needed to know the exact direction that

1. Or, at least, that is how astronomers look at it. To most other people, it is the stars that migrate systematically in the opposite direction behind the Sun.

points toward the shortest distance to the sacred Kaaba in Mecca, Saudi Arabia (not to be confused with Mecca, California, or Mecca, Indiana). It is this direction that Muslims use when it is time to pray toward Mecca. The solution is a nontrivial problem in spherical trigonometry that begins with a straight line that connects New York City to Mecca *through* Earth and then projects the line up to Earth's surface. The result is what is called a "great circle," which is normally the most desirous path for airplanes to fly. I computed the direction and told the caller. And like the proverbial boy scout who helps old ladies cross the street, I logged it as a public service deed for the day.

As you might expect, the annual path that the Sun appears to take against the background stars is obliquely tilted from the celestial equator at the same 23½-degree angle as the tilt of Earth's axis from a direction that is straight out of its plane of orbit. A solar or lunar eclipse can happen only when the Moon is very near the Sun's path. Reflecting this requirement, the Sun's path has been officially named the "ecliptic." The ecliptic and the celestial equator form tilted rings across the entire sky that intersect at two nodes. The angle of the tilt is called the "obliquity of the ecliptic," a decidedly mouth-filling phrase.

The Sun is south of the celestial equator for half the year and north of the celestial equator for the other half. Therein lies the origin of the variation in daylight throughout the year and the origin of the seasons. By definition, spring begins when the center of the Sun's disk crosses the celestial equator from south to north—the "ascending node." This is why newspapers report the particular minute of the day when spring begins. They could, if they felt so inclined, report the beginning of spring to the fraction of a second. By definition, summer begins when the Sun has climbed the farthest north of the celestial equator. This is where the two

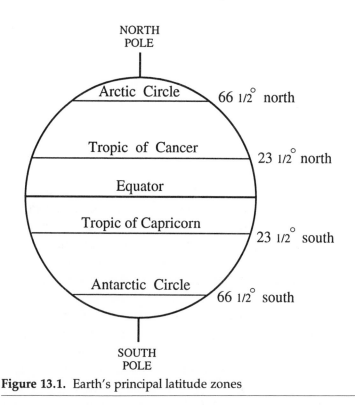

Figure 13.1. Earth's principal latitude zones

tilted rings have their greatest separation. As is true with spring, summer occurs at a particular moment that could be reported to the fraction of a second if there were public demand for such precision.

The important spots along the rest of the Sun's path can be readily deduced. The first moment of autumn is when the Sun crosses the celestial equator going south—the "descending node"—and the first moment of winter is when the Sun has descended the farthest south of the celestial equator before it resumes its journey northward. Two thousand years ago, on the first day of every summer, the Sun was superimposed on the constellation Cancer. The first day of summer is the only day of the year where the people on Earth who live at a latitude of 23½ degrees north get to have the noonday sun directly overhead. Not surprisingly, this band on Earth's surface can be identified on most maps and on all globes as the *Tropic of Cancer*. Equivalently, the first day of winter historically found the Sun to be superimposed on the constellation Capricorn. Only then can the residents along 23½ degrees south latitude enjoy a midday sun that is directly overhead. On Earth, this latitude is identified as the *Tropic of Capricorn*. At no time of any day in the year do Earth residents outside the region between 23½ degrees south and 23½ degrees north have a midday sun that is directly over-head. More bluntly stated, most of the population of the world has never seen the Sun directly overhead. They can only envy those who have traveled to the "tropics" or who just happen to live there.

The Sun begins its journey north along the ecliptic toward the celestial equator after the first day of winter. It begins to make larger and larger arcs across the daily sky, and thus stays in the sky longer and longer for Northern Hemisphere dwellers. If you have ever paid attention to the daytime sky, then you might have noticed that the winter sun rises far south of east and sets far south of west. The daily path is a low arc across the sky. In the summer, the Sun rises far north of east and sets far north of west. The daily path is a relatively high arc across the sky. During your lunch break, you can discover this for yourself if you measure the height of your shadow at noon on the first day of winter, and again at noon on the first day of summer.

A more revealing experiment, if you have nothing better to do for every one of your lunch breaks over the next year, is to stand in the same place every day at exactly 12 noon and put a mark on the ground where the top of your shadow

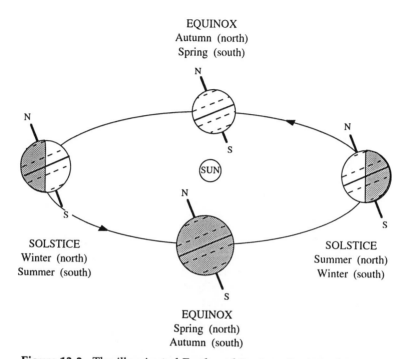

Figure 13.2. The illuminated Earth and its tipped axis is shown, in its orbit, at the first days of the four seasons. For the summer solstice in the north, the Tropic of Cancer receives direct sunlight at 12 noon. Notice that, at the same time, no part of the south polar cap will be rotated into sunlight; it remains in complete darkness. For the summer solstice in the south, it is the Tropic of Capricorn that receives direct sunlight at 12 noon, while at the same time, the entire northern Arctic cap remains in darkness. At the equinoxes, all parts of Earth, including the polar caps, receive exactly twelve hours of sunlight and twelve hours of darkness. Note: The Sun, and Earth's distance to it, are definitely not drawn to scale.

falls.[2] After a year of missed lunches, you will notice that your marks on the ground will grow longer and longer as December 21 approaches. The length of your shadow will pause for a day or two, and then by Christmas you will see it get shorter and shorter again for the six months up to June 21. Beginning June 21, your shadow's length will once again pause for a day or two before it begins to get longer and longer for the six months that lead back to December 21. You already know June 21 to be the first day of summer and December 21 to be the first day of winter. Your experiment showed that for each of these days the change in the length of your noon shadow stopped. If we deduce the Sun's behavior from your markings on the ground, we conclude that the noonday Sun reached its highest point on June 21 and its lowest point on December 21. In each case, before the Sun turned around, it appeared to stop for a day or two. This phenomenon is endowed with its own name: "solstice" (from the Latin *sol* for sun and *stitium* for stationary). The terms "summer solstice" and "winter solstice" are no less common than the "first day of summer" and the "first day of winter."

Had the descent of the Sun not stopped on December 21, then each day your shadow would continue to lengthen as the noon Sun gets lower. Eventually, the length of your shadow would become infinite—just before the noon Sun fails to appear above the horizon and you are abandoned in eternal darkness—one could make a horror movie about this. In the days of pagan rituals, the rebound of the Sun after December 21 was heralded as a joyous occasion. There were celebrations and festivities. When Christianity began to

2. Note that you cannot freeze your standing shadow in its place while you mark the ground. If your shadow behaves as it ought to, then it will follow you as you bend—so you may wish to solicit help from a friend. This shadow problem is a variant on the mirror problem, where your reflection does exactly what you do. The consequence: you can only kiss your reflection on the lips.

spread, and the uncertain birth date of Jesus Christ needed to be set, a time near this pagan Sun ritual (December 25) was selected to help promote the new religion with a minimum of resistance.

If you were extraordinarily precise during your year-long adventure in shadow etchings, so that each measurement was taken at exactly 12 noon, then you should notice that your marks on the ground have traced a figure "8." Because Earth's speed in its eccentric, oval-shaped orbit is not constant, and because the Sun seasonally finds itself above and below the celestial equator, the 24 hours of Earth's rotation does not always return the Sun to its highest spot on the sky. Sometimes the Sun gets there in a few minutes less than 24 hours, and other times it gets there in a few minutes more than 24 hours. This alternating speedy and tardy Sun is what causes the figure "8." On average, the Sun gets to its highest point in 24 hours, which is why household clocks needn't worry about such antics, even though sundials do. The figure "8" is also known as an "analemma," which occasionally makes a guest appearance—sideways and afloat—in the middle of the Pacific Ocean as drawn by globe makers. Perhaps there is no place else for them to put it.

The longer and shorter daytime arcs of the Sun are the cause of longer and shorter days. When I was a child, however, I was terribly confused. I knew that a solar day was always 24 hours, and that the rotation rate of Earth could be trusted, so I did not understand what people meant when they declared, "In the summer the days get longer." When I finally figured out that people were referring to the duration of daylight, I was still confused. Daylight hours begin to grow just after the first day of winter (the shortest day of the year). And they continue to grow through all of winter and all of spring until the first day of summer (the longest day of the year), at which time daylight begins to shorten again. So

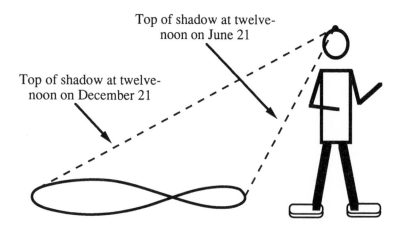

Figure 13.3. The stick figure with big feet is explaining that the length of your shadow at 12 noon on each day of the year will trace a figure "8." The tallest shadow will correspond to the low Sun during the winter solstice, near December 21. The shortest shadow will correspond to the high Sun during the summer solstice, near June 21. Because Earth's speed in its orbit is not constant, and because the Sun ambles above and below the celestial equator, the 24 hours of Earth's rotation does not always return the Sun to its highest spot in the sky. The Sun can be "fast" or "slow," which traces an "8" over the year. This famous figure "8" is known as an "analemma."

let it be known among the confused children of the land that *winter* is the season where days get longer and *summer* is the season where days get shorter. Perhaps British children are less likely to get confused since the first day of summer in the United Kingdom is called "midsummer," and the first day of winter is called "midwinter."

On the first day of spring and of autumn, the Sun crosses the celestial equator. These are the only days of the year where every Earth resident experiences daylight of equal duration to the night. These two days are more commonly

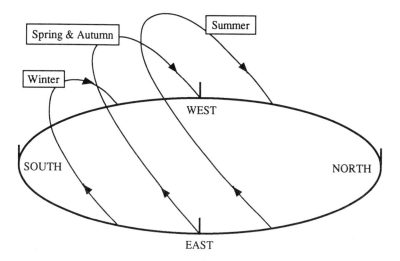

Figure 13.4. Path of the Sun through the daytime sky on the first day of each season when viewed from an intermediate latitude in the Northern Hemisphere, which is appropriate for most of the world's population. Notice that only on the first day of spring and of autumn does the Sun rise due east and set due west. The low arc of the Sun in December and surrounding months provides much less than twelve hours of light, while in June and its surrounding months the high arc provides much more than twelve hours of light. Only near the spring and autumn equinoxes does the duration of daylight and the duration of night each equal twelve hours.

called the vernal (spring) and the autumnal (autumn) "equinox," from the Latin *æqui* for equal and *noct* for night.

The lengthening of the daytime hours from winter to spring is accompanied by sunlight that is more direct, and consequently more intense, on Earth's surface. The slow and continued day-to-day increase in sunlight heats the hemisphere as the season changes from winter to spring. At any moment of the year, the opposite transition is happening in the Southern Hemisphere. What does this say for the equa-

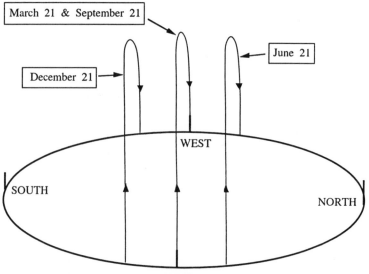

Figure 13.5. Path of the Sun through the daytime sky for equatorial residents. Only along the equator does every day contain twelve hours of light and twelve hours of darkness. There are no seasons and nobody hibernates.

tor? Being caught exactly in the middle, its residents experience no seasons. On the equator, every day is equivalent to an equinox. There are also no deciduous trees, no hibernating animals, and no canceled school days from snowstorms. What does this say about the poles? Beginning at 66½ degrees latitude (which, by the way, is the 90-degree latitude of the pole minus the 23½-degree tilt of Earth's axis) and heading toward the pole, there will always be at least one day where the arc of the Sun is so broad that it is, in effect, broader than the entire horizon, with the result that the Sun does not set. The 66½-degree north latitude is unimaginatively called the "Arctic Circle" while the 66½-degree south

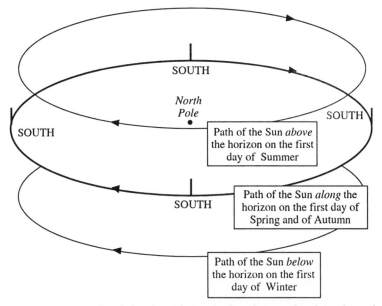

Figure 13.6. Path of the Sun through the sky on the first day of each of the four seasons as viewed from the North Pole. In the summer the Sun never sets. A "day" lasts six months while the Sun's path spirals slowly toward the horizon. Near the equinoxes, the Sun's disk remains half up and half down along the horizon as sunset (or sunrise) lasts for many days. The winter path of the Sun remains completely below the horizon for six months. If you travel in any direction from the North Pole, you can only head south. A diagram for the South Pole would appear identical, except that the arrows would point in the opposite direction, and all roads would lead north.

latitude is called the "Antarctic Circle." Nearer and nearer to the poles, the number of days in a year grows for which the Sun does not set. This event is known as the "midnight sun" in many places, but they could just as accurately, though less romantically, call it the "11:30 P.M. Sun" or the "1 A.M. Sun." By the time you get to the poles, you will notice that the Sun rises just once a year and, of course, sets just once a year. The consequence: a six-month day and a six-month night.

I once received a telephone call from an orthodox Jew who was planning a summer trip to Alaska. He needed to know the exact time of the setting Sun for the Fridays of his trip, which signals the onset of the Jewish Sabbath.[3] I told him he had better keep out of the Arctic Circle, and I gave the caller sunset times for more southern latitudes in Alaska.

From the point of view of an observer perched "above" the solar system, summer in the Northern Hemisphere is where the north pole of Earth's axis is tipped toward the Sun. Six months later, with Earth on the other side of the Sun, the same tilt of the axis now points away from the Sun. As noted in chapter 12, just as a spinning and tilted top will wobble, so does the spinning and tilted Earth. Since a full wobble takes about 25,700 years to complete, you need not worry about getting tossed off the surface of the Earth. One of several cosmic consequences is that one-half a wobble from now (the year A.D. 15000) Earth will be tipped the other way. Polaris, the North Star, will become Polaris, the ex-North Star. The constellations that are normally identified with the nighttime winter sky will have shifted to become summer constellations, and the summer constellations will have shifted to become visible in the winter. In other words, the celestial grid—complete with its celestial equator, the path of the Sun, their nodes of intersection, and the celestial poles—will be projected onto a backdrop of stars that is offset from before.

Indeed, Earth has wobbled enough already so that the position of the Sun against the backdrop of stars on the first

3. The Jewish Sabbath lasts from sunset Friday to sunset Saturday.

day of summer no longer falls in the constellation Cancer—so that the name "Tropic of Cancer" is technically no longer appropriate. The current backdrop is the constellation Gemini. Additionally, the Sun on the first day of winter now has the constellation Sagittarius as a backdrop, not Capricorn—so that the name "Tropic of Capricorn" is also no longer appropriate. Either by tradition or perhaps a mandate from frustrated map and globe makers, the Tropic of Cancer and the Tropic of Capricorn have retained their names in spite of this early-breaking news. Two thousand years from now, perhaps you can lobby the mapmakers to introduce the names of the next relevant constellations: the *Tropic of Taurus* and the *Tropic of Scorpius.*

After its 500-second journey, light from the Sun must cross from the vacuum of interplanetary space to Earth's atmosphere. Upon traversing the boundary between these two regions of different density, the speed of light will drop, which beacons an underunappreciated fact of physics: the speed of light through anything other than a vacuum will always be less than it is in a vacuum. When light penetrates at oblique angles, then the direction of motion changes as well. This phenomenon is known as refraction and is the principle that allows lenses, and of course eyeballs, to focus light. The deeper into Earth's atmosphere the light travels, the more it refracts as the atmosphere gets denser an denser. What all this means is that the Sun is not where you think it is in the sky. At sunset, as our precious orb of glowing hydrogen poses prettily upon the horizon, the refraction of its light is greatest. Indeed, the unrefracted Sun has already set. Don't tell your lover, but every romantic memory of a sunset (or sunrise) in your life is the consequence of a refracted image of the Sun, and not the Sun itself. Of course, the same is true for the Moon since its light also originates from outside of Earth's atmosphere. The song that contains the lyric "It's only a paper moon" could easily be reworded to "It's only a refracted image"—with no loss of relevance to the song's content.

People who go fishing with a bow and arrow know all about refraction. Do not aim where you see the fish—you will miss. The fish you see is a refracted image formed as the light from the real fish bends upon crossing the boundary from water to air. Those who are experienced know that to nab the fish you must aim at the correct angle beneath it. In honor of this talent, maybe people who fish with a bow and arrow should be called "anglers."

Moon

Earth's moon holds a special place in my heart. It was a view of the first quarter Moon (the phase that many people call "half") through binoculars at age eleven that triggered my career path to study the universe. The mountains and valleys and craters were revealed in detail that I could not have imagined from a simple glance with the unaided eye. With greater academic sophistication, I soon began to appreciate other aspects of the Moon that are just as ogle-worthy: (1) the Moon is the only satellite in the solar system that has no name; (2) the Moon is in predictable gravitational orbit around Earth; (3) the orbit of the Moon sometimes gets in the way of our view of the Sun, which spawns one of nature's greatest spectacles—a total solar eclipse; (4) on occasion, the Moon ambles into Earth's shadow, which extends nearly a million miles into space, and spawns yet another spectacle—a total lunar eclipse; (5) the Moon is in a gravitational "tidal lock" with Earth, which prevents the far side of the Moon from ever facing Earth; and (6) the Moon is made of rocks and not some variety of smelly exotic cheese.

You can actually observe the Moon's motion in orbit

around Earth, even though it is about as exciting as watching the hour hand on a clock. The next time you spot the Moon at night, take notice of the pattern of stars that surround it and of the Moon's position relative to them. Go back inside for about three hours and then return to see the Moon. You will see that it moved east relative to the background stars by an amount equal to its own diameter. The cumulative effect of this daily orbital motion is for the Moon to rise about 52 minutes later and set about 52 minutes earlier each day. This slow, steady, and systematic motion continuously changes our view of the illuminated Moon relative to the Sun. We see the Moon "wax" (grow) from a thin crescent, which sets shortly after the Sun, to a first quarter, commonly known as a "half moon," which sets at about midnight. The Moon phase continues to wax until it is full. Full moons rise just after sunset and set just before sunrise. The portion of the Moon's illuminated surface that faces Earth next wanes to last quarter, which sets at 12 noon, and then to crescent, which sets just before sunset. The phase between the waning and waxing crescents is called the "new moon." It is the only unobservable phase because the entire far side of the Moon receives complete illumination.

In a clash of terminology, I once received a telephone call from someone who wanted to know when the next new moon was to occur. This is, of course, a single moment in time as the Moon passes between the Sun and Earth. I gave the caller the information, but then the caller asked when this new moon would be visible from New York City. I knew, at the time, that Ramadan was near. This is the ninth month of the Muslim calendar that is traditionally a period of daily fasting—it begins and ends with the sighting of what is called the new moon. But what the Muslims—and almost any other religious or social culture—refer to as the "first sighting of the new moon" is actually the first sighting of the waxing crescent in the early evening sky toward the west, just after sunset. For this to happen, the Moon must emerge from its new phase to be far enough away from the Sun in the sky so that you obtain a crescent-shaped glimpse of the illuminated half. This normally takes a day or two beyond the new moon.

The phases of the Moon (as well as tons of other informa-

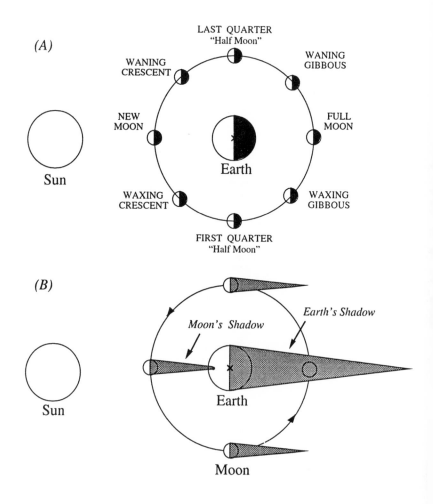

tion) are tabulated in a book called the *Astronomical Almanac*, formerly the *Astronomical Ephemeris and Nautical Almanac*, which is published annually by the nautical almanac offices of the United States Naval Observatory in Washington, D.C., and of the Royal Greenwich Observatory in Greenwich, England. The word "almanac" also appears in the title of the annually published reference, *The Old Farmer's Almanac*, where weather predictions were traditionally made from a secret formula—devised by the founder—which is contained in a black tin box located in Dublin, New Hampshire. One particular occasion, a caller to my office wanted to plan a honeymoon vacation around the full moon. When I told the caller that my source for the Moon's phases is the *Astronomical Almanac*, the response was, "Then predicting the phase of the Moon must be like predicting the weather, you really cannot know for sure what it will be the next day." I did not know whether to compliment the caller on such healthy skepticism of the weather predictions from *The Old Farmer's Almanac* or whether to chide the caller for never having noticed the daily, predictable changes of the Moon.

Figure 13.7. The Sun, Moon, and Earth. *(A)* the Moon's orbit around Earth as it would appear if we were to look down on the North Pole. Note: The Sun and relative distances are definitely not drawn to scale. Half the Moon is always illuminated by the Sun. The part of this illuminated surface that happens to fall inside the orbit is what is visible to people on Earth. This Ferris-wheel of angles is the origin of the Moon's phases. *(B)* On occasion (once every couple of years), the tip of the dark cone of the Moon's shadow sweeps across Earth's surface during new moon. Only the people who happened to live in the shadow's path, or who are rich enough to travel to it, will actually see a total eclipse. A half month later, when the Moon's orbit takes it to the other side of Earth, the Moon will enter Earth's shadow and display a total lunar eclipse, which is visible for everyone on the nighttime side of Earth.

Actually, I did both and then explained that, with the exception of rare typographical errors, the *Astronomical Almanac* is 100 percent correct, every day of every year. And that it contains no horoscopes, folk remedies, or cute human-interest stories.

For many people in the world, the rising full moon is one of the top wonders of nature—especially if the horizon is dotted with trees or buildings as the Moon emerges from behind. This wonderment often includes a full case of the "Moon on horizon illusion," where the orb appears unnaturally large as it rises or sets. While there is still no agreement among Moon-on-horizon experts, it is almost certainly related to a confusion in your depth perception induced by familiar objects on your horizon. A full moon, and the presence of identifiable buildings or trees, adds considerably to the illusion. Sales brochures for romantic cruises notwithstanding, moonrise over an expanse of ocean—where there are few horizon depth cues—provides a relatively poor moon-on-horizon moment. It is rumored that if you observe the rising moon through your legs while bent over, then the moon-on-horizon illusion will also be significantly lessened because the trees and buildings are no longer registered as recognizable icons. Feel free to attempt the experiment when nobody is looking.

The human fascination with the Moon on the horizon is powerful. On yet another occasion, I once received a phone call from a cinematographer of a film in production by Francis Ford Coppola. The cinematographer wanted to obtain genuine footage of the full moon as it rose over the Manhattan skyline. The film clip would be spliced into the film to establish the urban "night mood." I was asked to provide the best time, date, and location for this task. Only after the telephone call did it occur to me that the full moon's photogeneity is what gets it artificially selected for appearances in feature films. The other moon phases, which are also cosmically legitimate, tend to be neglected.

I was also concerned that Coppola's clip was going to feed the misconception that the Moon only comes out at night. Please tell your friends that the Moon is visible in broad daylight on about twenty-four of the 29½-day cycle of phases. The film clip may also feed the idea that the full moon is common. But the Moon spends ten days of its 29½-day cycle being a crescent and another ten days being that funny-looking intermediate phase between quarter and full, which is officially called "gibbous."

Perhaps I am biased, but nights with full moons are the most avoided nights of the year among the world's professional astronomers. The full moon is so bright (it is over five times brighter than the combined light of two side-illuminated "half" moons) that the number of detectable objects in the night sky drops precipitously. The full moon is not even interesting through binoculars. Being front illuminated as seen from Earth, the Moon has no shadows among its mountains, hills, and valleys that would otherwise reveal surface texture and depth. A professional portrait photographer would never illuminate someone from directly in front because the person's face would then look flat, dull, and lifeless. Lights are typically placed at some oblique angle to provide shadows among the facial features. Although, if the person has a serious case of acne, then detailed facial texture may not be what is sought.

It is not fully appreciated that the Apollo astronauts on the Moon's surface could always communicate with mission control. As seen from the near side of the Moon, Earth is always in the sky, which can only be true if the Moon rotates on its axis in exactly the same amount of time that it takes for the Moon to orbit Earth. Indeed, the Moon is in a "tidal lock" with Earth such that it always shows the same face.

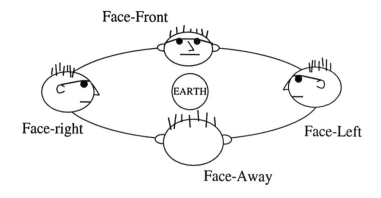

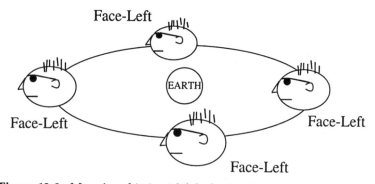

Figure 13.8. Moon's orbit in tidal lock. In the upper panel, the head must rotate continuously in its orbit to show the same face toward Earth. As is also true for the real Earth and Moon, there is a perfect match between the rotation period and the time for one complete orbit. This "tidal lock" ensures that observers on Earth will never see the far side of the Moon. If the head did not rotate, as pictured in the lower panel's set of orbs, then it would always face the same direction and observers on Earth would get to view all sides of the head during a full orbit.

Yes, there is a near side and a far side of the Moon, but since all parts of the Moon receive sunlight at different times in its monthly orbit, *there is actually no such concept as the dark side of the Moon*. It may require a century of effort among astronomy educators to undo the influence of the popular rock group Pink Floyd, whose 1973 album entitled "The Dark Side of the Moon" misled an entire generation of Americans.

The Earth-Moon tidal lock is not a cosmic coincidence. It is the natural consequence of strong tidal forces on a nearby rotating object. A similar condition exists for the large planets (Jupiter, Saturn, Uranus, and Neptune) with their inner satellites and for the Sun with Mercury. The Moon's tidal forces are at work on Earth, which, among other things, act to slow Earth's rotation rate. Eventually, the rotation rate of Earth itself will equal the time it will take for the Moon to complete one orbit. The result: Earth will show only one face toward the Moon the way the Moon shows only one face toward Earth. This will take several hundred billion years, so you needn't worry about it just yet. In the meantime, you can "watch" it happen as the occasional leap seconds are introduced to the calendar year by the International Earth Rotation Service.

Eclipses

The Moon's orbit around Earth is tipped about five degrees from the path of the Sun against the background stars. As a consequence, the Moon crosses the ecliptic twice for each complete orbit. If the Moon's phase is new when it crosses the ecliptic, then Earth, Moon, and Sun are aligned in syzygy, and earthlings are treated to a total solar eclipse. No, not all earthlings. Just the ones who are lucky enough to have the Moon's narrow shadow pass over their town, or the ones who are rich enough to travel to the shadow's path.

The dark cone of the Moon's shadow, the umbra, just barely reaches Earth in a fast-moving dark circle that is typically one hundred miles wide. The range among eclipses extends from zero to about two hundred miles. In what would otherwise be broad daylight, the Sun disappears behind the Moon. Strictly speaking, any time one cosmic object passes in front of another, as in a total solar eclipse, the event is known as an "occultation."

On Earth, the Moon and Sun appear roughly the same size in the sky. They are each about one-half degree in angle. An excellent protractor, for those emergencies when you must measure an angle in the sky, is your fist at arms length. It spans about ten degrees for the average human. If you align the bottom of your fist with the horizon, then nine fists (your left and right fist alternatively stacked) should leave you straight overhead at a 90-degree angle from where you started. If you have big fists, then you probably also have long arms, which ensures that your fist still spans ten degrees at arms length. (For this method to fail, you would need the arm-to-fist proportions of an orangutan.) At one-half degree, the Sun and Moon each spans less than one-fourth the width of your finger at arms length.

The near-match in angular size between the Sun and Moon allows the outer atmosphere of the Sun, known by the poetic term "corona," to be revealed during the few minutes of totality. If you know which way the Moon's umbra will approach, then a glance toward the horizon in that direction during the few seconds before totality will reveal a fast-moving column of darkness that looks as though the sky were being parted. In the precious few minutes of totality, the entire sky darkens, the stars become visible, the solar corona glows with gentle radiance, the air temperature drops, and animals behave strangely—especially humans. Humans temporarily leave their job to spend

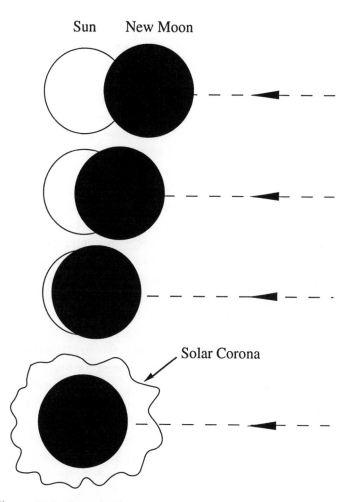

Figure 13.9. A total solar eclipse, where the Moon is large enough to cover the Sun completely. As drawn, the Moon moves in its orbit around Earth from right to left.

wads of money traveling to exotic spots on Earth's surface via car, plane, and ocean liner. They spend millions of dollars on eclipse memorabilia. And they suffer great mental trauma if clouds appear on the day of the eclipse.

I was one of these strangely behaving humans when I saw the seven-minute total solar eclipse of June 30, 1973—one of the longest on record, with a moon shadow on Earth that was 185 miles wide. I was on board a large ocean liner that sailed into the path of the Moon's shadow in the Atlantic Ocean, off the coast of northwest Africa. Ocean liners give you the option to sail to a spot with a good weather forecast so I did not risk mental trauma. There was one woman on the ship, however, who *did not* act strangely. What was shocking about her behavior was that she seemed to function in an alternative reality—only by not acting strangely did her behavior look strange. During totality, everybody else on the ship (myself included) rattled off dozens of photographs while grunting assorted primitive syllables such as *ooooh* and *aaahhhh*. Meanwhile, in a vision equally as surreal as the total eclipse, this woman was knitting a sweater while comfortably seated on a deck chair. This was my first lesson that perhaps the marvels of universe do not induce awe in everyone.

The eccentric orbit of the Moon around Earth brings it within 220,000 miles and takes it as far away as 255,000 miles. Similarly, the eccentric orbit of Earth around the Sun brings it as close as 91,500,000 miles and as far as 94,500,000 miles. The apparent size of the Sun and the Moon in the sky changes accordingly. There are some solar eclipses where not only is the Earth-Moon distance larger than average, but the Earth-Sun distance is smaller than average. Under these circumstances, the dark cone of the Moon's shadow does not reach Earth's surface. From Earth's point of view, the Moon's size in the sky is not large enough to cover the Sun

Sun   New Moon

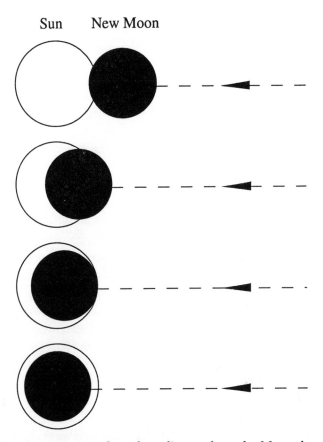

Figure 13.10. An annular solar eclipse, where the Moon does not happen to be big enough in the sky to cover the Sun completely. The Sun appears to enclose the Moon the way an amoeba encloses its dinner. As drawn, the Moon moves in its orbit around Earth from right to left.

completely, so that as the eclipse proceeds, a ring of sunlight encloses the Moon the way a hungry amoeba encloses its dinner. These eclipses have been dubbed "annular eclipses" for the annulus of sunlight that remains during mid-eclipse.

During all solar eclipses, the Moon's shadow blazes across Earth's surface between two and three thousand miles per hour—it will most certainly outrun you. As lyrical as it may otherwise sound, you will never be casually *followed* by a moon shadow.

If the Moon's phase is full when it crosses the ecliptic, then once again Earth, Moon, and the Sun are in syzygy, but earthlings are now treated to a total lunar eclipse. The Moon, in its orbit, crosses the 850,000-mile-long shadow cone of Earth's umbra. At the distance to the Moon, Earth's umbra is over three times as wide as the full moon, so the entire eclipse takes many hours. An unsuspecting glance at the eclipse in progress looks as though the Moon had spontaneously decided to cycle through phases, with Earth's umbra taking bigger and bigger bites. During totality, when the Moon has completely entered Earth's umbra, the Moon all but disappears without much spectacle or fanfare. Unlike the narrow path of a total solar eclipse, nearly everyone on the same side of Earth as the full moon will bear witness to a lunar eclipse. So while they are not more common than solar eclipses, far more people get to view lunar eclipses from their own backyard or roof. Compared with total solar eclipses, total lunar eclipses are long and, quite frankly, boring.

Nights during or near the full moon (known as "bright time" by astronomers) are the *least* desirable nights to observe the universe because the sky is hopelessly contaminated with moonlight. To the unaided eye, the number of detectable stars drops from over three thousand during new moon to about three hundred during full moon. And nebulous extended objects such as galaxies are decidedly less impressive. Nearly all discoveries of dim galaxies at the edge of the universe have occurred during or near new moon (known as "dark time" by astronomers) at the world's major observatories. On May 25, 1975, there was a total lunar eclipse, which a group of astronomers at the California Institute of Technology in Pasadena, California, deemed enough of an excuse to hold an evening party. When the eclipse began, it was noticed that a particular astronomer did not show up for the gathering. One of those in attendance recalled that the missing astronomer had suspiciously requested time on the 200-inch Palomar telescope during the full moon to observe a very dim object. By mid-eclipse it simultaneously occurred to all assembled that the missing astronomer was clever enough to request observing time during the full moon—knowing that it was to be eclipsed and knowing that the observing conditions during a totally eclipsed full moon rival the darkest skies of a new moon.

Planets

If there exists a cosmic ballet, it is among the solar system's planets as they wander against the background stars with orbits and paths that are choreographed by the forces of gravity. With an occasional cameo appearance by the Moon, the planets (especially the five visible to the unaided eye: Mercury, Venus, Mars, Saturn, and Jupiter) assemble in different combinations at different times of the year to create striking photo opportunities. The planets, in their orbits, have enchanted stargazers for centuries. In the days before computer simulations, people even built orreries, which are mechanical working models of the solar system. They served as a teaching tool and as toys to play with on a cloudy nights.

As noted in chapter 12, all planets in the solar system

orbit the Sun in roughly the same plane. The observational consequence is that the ecliptic is shared by all other planets. It is a veritable planetary freeway of the sky. Perhaps it should, instead, be called a *high*way. One should expect many occasions each year where several of these objects are found in the same region of the sky. Indeed, when two or more objects can fit within the field of view of ordinary binoculars, then we say they are in "conjunction." In my opinion, which many share, the most photogenic conjunctions occur when one or more planets assemble with the crescent moon against the deeply colored curtain of the twilight sky. This can happen during dusk with the waxing crescent moon, or as those who work the "graveyard shift" know, it can happen during the early dawn with the Moon as a waning crescent.

If Earth's lower atmosphere is more turbulent than usual, then the path of starlight becomes severely disrupted as it refracts unpredictably across the different air densities. When this happens, stars begin to "twinkle." When atmospheric turbulence really gets bad, even planets will twinkle. All this twinkling may sound poetic and may look pretty during a conjunction, but it represents the *worst possible seeing conditions that an astronomer can encounter.* (Actually, total cloud-cover is slightly worse.) This dilemma was documented in 1704 by Sir Isaac Newton in his seminal treatise on optics.

> If the Theory of making Telescopes could at length be fully brought into Practice, yet there would be certain Bounds beyond which Telescopes could not perform. For the Air through which we look upon the Stars, is in a perpetual Tremor; as may be seen by the . . . twinkling of the fix'd Stars.

Sir Isaac continued with telescopic foresight:

The only Remedy is a most serene and quiet Air, such as may perhaps be found on the tops of the highest Mountains above the Grosser Clouds.[4]

The well-publicized Hubble Space Telescope was lifted into orbit primarily to escape the degraded image quality and poor resolution that the lower atmosphere imposes on observations of all objects.

Arguably, the world's most famous painting that portrays stars is "The Starry Night" by the nineteenth-century Dutch impressionist Vincent van Gogh. These stars are drawn as large circular undulating yellow-white blobs in the sky. If this is what Vincent actually witnessed, assuming his eyeballs did not suffer from a bad case of astigmatism, then it must go down in the annals of astronomy as the worst seeing conditions ever recorded for a clear night.

In my early years of high school, I attended a summer camp for kids who knew they wanted to grow up to become astronomers. It was located in the cloudless skies of the Mojave Desert of southern California where we lived nocturnally for two months. The camp was equipped with a bank of more than a dozen telescopes of various sizes, each designed for a particular scientific purpose. A friend of mine at the camp received a letter from home that said all the usual tender things that letters from home say. Except that the letter ended with an unwittingly declared curse from Hell:

. . . and we hope that all your stars are twinkling!

Love,
Mom & Dad

Sometimes a twinkling planet in the twilight sky can be quite striking, especially if it is Venus. Because of its proxim-

4. Sir Isaac Newton, *Opticks* (Book 1, Part 1, Proposition 8, Problem 2; rpt., New York: Dover, 1979 [1704]).

ity to Earth, and because of its high albedo from a thick reflective white cloud-cover, Venus is often the brightest object in the sky. At its brightest it is nearly twenty times brighter than Sirius, the brightest star in the nighttime sky. When Venus is low on the horizon, a turbulent atmosphere can sometimes behave like a prism and display quite a show of twinkling colors. For these reasons, Venus is occasionally mistaken for an unidentified flying object that hovers over the horizon. For some people, a UFO means a flying saucer that is commanded by hostile aliens. To other people, a UFO is simply an object that they cannot identify. In general, it is safer to admit uncertainty and to inquire further than it is to invoke extraordinary imagination—particularly if you are otherwise unfamiliar with that evening's schedule of cosmic conjunctions.

For example, in some urban settings the sky is unfamiliar to many people. I submitted the following recollection to the *New York Times,* which was printed in their "Metropolitan Diary" of Wednesday, July 12, 1991.

Dear Metropolitan Diary,

An elderly sounding woman with a strong Brooklyn accent recently called my office at Columbia University's Department of Astronomy to ask about a bright glowing object she saw "hovering" outside her window the night before. I knew that the planet Venus happened to be bright and well-placed in the west for viewing in the early evening sky, but I asked more questions to verify my suspicions. After sifting through answers like, "It's a little bit higher than the roof of Marty's Deli," I concluded that the brightness, compass direction, elevation above the horizon, and time of observation were indeed consistent with her having seen the planet Venus.

Realizing that she has probably lived in Brooklyn most

of her life, I asked her why she called now and not at any of the hundreds of other times that Venus was bright over the western horizon. She replied, "I've never noticed it before." You must understand that to an astronomer this is an astonishing statement. I was compelled to explore her response further. I asked how long she has lived in her apartment. "Thirty years." I asked her whether she has ever looked out her window before. "I used to always keep my curtains closed, but now I keep them open." Naturally, I then asked her why she now keeps her curtains open. "There used to be a tall apartment building outside my window but they tore it down. Now I can see the sky and it is beautiful."

The path of the planets through the sky is not as simple as that of the Moon or the Sun. Yes, the planets orbit the Sun. And yes, if you looked from night to night you would see them move against the background stars. But what complicates this simple picture is that we observe planets that orbit the Sun while riding on a planet that orbits the Sun. The resulting planetary paths confounded centuries of the world's greatest thinkers before there was agreement that the Sun was the center of planetary motion.

All planets orbit counterclockwise[5] when viewed from "above" the Sun. When viewed from Earth, a general trend emerges for planets to move from west to east against the background stars. The inner two planets (Mercury and Venus), complete their orbits around the Sun faster than Earth. The outer planets, however (Mars, Jupiter, Saturn, Uranus, Neptune, and Pluto), take longer than Earth to complete their orbits around the Sun. A simple and direct obser-

5. During your life, if all your clocks had digital faces, then "counterclockwise" is the direction that baseball players, track runners, horses, and race cars move around their respective tracks.

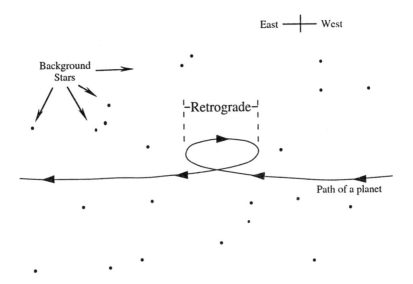

Figure 13.11. The west-to-east general path of a planet can be seen in the sky against the background of stars. As Earth and the planets orbit the Sun, there will always be an interval where the path of a planet will appear to loop backwards (east to west) in what is called "retrograde" motion.

vational consequence is that there will always arrive a time interval when the relative motion between Earth and each of the other planets makes them appear to move in "reverse," from east to west, against the background stars. If you do not put the Sun at the center of planetary motion, you will have an extraordinarily difficult time explaining what you see. In spite of this, the historical bias toward an earth-centric view of the universe was strong. When the sixteenth-century Polish astronomer Nicolaus Copernicus wrote *De Revolutionibus* (a treatise that placed the Sun, rather than Earth, at the center of planetary motion), an anonymous foreword was inserted at the time of publication without

Copernicus' knowledge or permission. It was later revealed to have been written by the Lutheran theologian Andreas Osiander, who had helped to supervise the printing. The foreword included the following disclaimer:

> To the reader Concerning the Hypothesis of this Work
>
> There have already been widespread reports about the novel hypotheses of this work, which declares that earth moves whereas the sun is at rest in the center of the universe. . . . For these hypotheses need not be true or even probable. On the contrary, if they provide a calculus consistent with the observations, that alone is enough.

The concept of backward apparent motion should be easy for modern humans. The next time you visit an amusement park, give close attention to the dizzy people on the rides that go in circles. (Ignore the people doing energy experiments on the roller coaster.) In an analogous scenario to orbiting planets, you will notice that when the riders are near you on these nausea-inducing machines they might cross your field of view from left to right, yet when they are on the other side of the machine the reverse is true—you will see them pass from right to left. Similarly, these people see you, as you wait patiently in line for the next ride, shift across their field of view alternatively from left to right and then from right to left.

Planets that appear to move backward are commonly said to be in "retrograde," which has even found its way into Shakespearean literature. In the first scene of the first act of the comedy *All's Well That Ends Well,* Helena displays a sharpness of wit as she comments on the valor of Parolles.

HELENA: Monsieur Parolles, you were born under a charitable star.
PAROLLES: Under Mars, I.

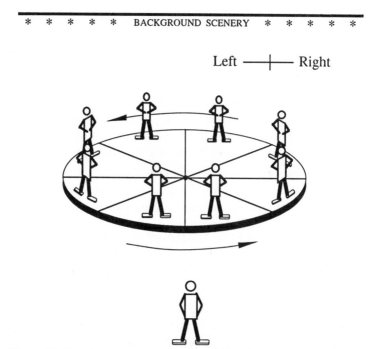

Figure 13.12. A person who waits to join others on an amusement park ride will notice that the revolving people who feel nausea alternatively pass left to right and right to left against the background scenery. To the people on the ride, the person who waits will also appear to swing left to right and then right to left. Planets in orbit around the Sun exhibit similar behavior when viewed from Earth.

HELENA: I especially think, under Mars.

PAROLLES: Why under Mars?

HELENA: The wars hath so kept you under that you must needs be born under Mars.

PAROLLES: When he was predominant.

HELENA: When he was retrograde, I think rather.

PAROLLES: Why think you so?
HELENA: You go so much backward when you fight.

Unlike amusement park rides and Shakespeare's Parolles, planets require months of careful tracking to watch them enter and emerge from retrograde motion against the background stars. The observation is a task best accomplished by astronomers and insomniacs.

Of all the cosmic objects that one might observe with a backyard telescope, the planet Saturn, with its banded surface, its orbiting moons, and its awesome ring—parted in its middle by Cassini's division—would appear high on the list for its ability to excite passers-by. In my youth, I did not have a backyard, only the roof of my urban apartment building. And there were no passers-by, except for the occasional grumpy police officer who would mistake my telescope for an M-79 grenade launcher. My telescope's motor, which allowed the telescope to track stars across the sky as Earth rotates, required electricity. I would often lower a 100-foot extension cord from the roof through my bedroom window, which police would reliably mistake for a rappelling rope. I had a total of five such encounters. In three of the five cases, I was promptly saved by the planet Saturn, with the following a typical dialogue:

OFFICER (*shooting-hand poised near gun, other hand holding flashlight*): What the hell is that thing, and what are you doing on the roof?
ME (*maneuvering Saturn quickly into field of view*): Good evening, officer. Ever see the planet Saturn through a telescope before?
OFFICER (*shooting-hand now scratching head*): No, just in pictures.
ME: Turn off your flashlight and have a look.

OFFICER (*looking through telescope*): Wow! Saturn really does have rings! Maybe I'll buy one of these for my kids!

The police officers may have learned that in life and in the universe, it is always best to keep looking up. But if somebody really does set up a rooftop grenade launcher, I hope it will still attract their attention.

Suggested Reading

Periodicals

The following periodicals will keep you abreast of discoveries in astronomy and science in general.

Scientific American, 415 Madison Avenue, New York, New York 10017. Available on most newsstands. With articles written by research scientists, it is still the most respected science periodical intended for the lay reader.

Science News, 1719 N Street, N.W., Washington, D.C. 20036. Weekly news items on discoveries in all areas of scientific research. Highly readable.

Sky and Telescope, Sky Publishing Company, 49 Bay State Road, Cambridge, Massachusetts 02238. Indispensable if you own a telescope and are serious about getting around the sky.

Astronomy Magazine, P.O. Box 92788, Milwaukee, Wisconsin 53202. Looks better on the coffee table than *Sky & Telescope* but contains mostly the same information. Its photogenic layout attracts a considerably larger audience.

Stardate Magazine, McDonald Observatory, University of Texas, Austin, Texas 78712. Many of the chapters contained in *Universe Down to Earth* first appeared as feature articles in this magazine. Published six times a year, its articles, news items, star charts, and question-and-answer column are ideally suited for the weekend stargazer. *Stardate Magazine* is published by the same people who produce the nationally syndicated daily radio spot "Star Date."

Griffith Observer, 2800 East Observatory Road, Los Angeles, California 90027. Serves a similar audience to *Stardate Magazine*. They are each for people who cannot get excited about all the contents of *Sky and Telescope,* but who care enough about the universe to stay in touch with what is important.

Skeptical Inquirer, The Quarterly Journal of the Committee for the Scientific Investigation of the Claims of the Paranormal, Box 703, Buffalo, New York 14226-0703. Perhaps the most important periodical in this list, the *Skeptical Inquirer* applies scientific principles and methods of investigation to the extraordinary claims that permeate the sensationalist press. The topics extend from UFO abductions to telekinesis to out-of-body experiences. The magazine is society's vaccine against pseudoscience.

Books

Many of the topics discussed in *Universe Down to Earth* may be explored further in the following books.

Biography and Autobiography

Christianson, Gale E. *In the Presence of the Creator: Isaac Newton and His Times.* New York: Free Press, 1984. A fat biography that leaves no apple unturned.

Einstein, Albert. *Out of My Later Years.* New York: Philosophical Library, 1950. Einstein reflects on his life as a scientist and as a public icon.

Geymonat, Ludivico. *Galileo Galilei: A Biography and Inquiry into His Philosophy of Science.* Translated by Stillman Drake. New York: McGraw-Hill, 1965. The life and times of Galileo.

Koestler, Arthur. *The Sleepwalkers.* New York: Penguin, 1990. A chatty and often amusing account of some great astronomers of the past.

Payne-Gaposchkin, Cecilia. *An Autobiography, and Other Recollections.* Edited by K. Haramundanis. New York: Cambridge University Press, 1990. It was not easy being a female astronomer in the early twentieth century.

General Astronomy

Adouze, Jean, and Guy Israel, eds. *Cambridge Atlas of Astronomy.* 2d ed. New York: Cambridge University Press, 1988. If you seek to own only one astronomy book (besides, of course, *Universe Down to Earth*), it should be this one. With contributions from professionals in the field, it is a comprehensive reference on all areas of astronomy.

Burke, John G. *Cosmic Debris: Meteorites in History.* Berkeley: University of California Press, 1986. If you are one of those people who maintains a morbid fascination with meteors that strike Earth, then this book and the following Chapman/Morrison title are for you.

Chapman, C. P., and D. Morrison. *Cosmic Catastrophes.* New York: Plenum, 1989.

Harwit, Martin. *Cosmic Discovery: The Search, Scope, and Heritage of Astronomy.* New York: Basic Books, 1981. A telling account of the academic and societal conditions that have led to discoveries in modern astronomy.

Jaschek, Carlos, and Mercedes Jaschek. *The Classification of Stars.* New York: Cambridge University Press, 1987. A confusing topic, even for many professional astronomers, this book explains it all.

Moore, Patrick. *Fireside Astronomy: An Anecdotal Tour Through the History and Lore of Astronomy.* New York: Wiley, 1992. An anthology of unusual, bizarre, and humorous recollections from one of the most prolific British astronomers of our times.

Shu, Frank H. *The Physical Universe: An Introduction to Astronomy.* Mill Valley, California: University Science Books, 1982. If simply knowing the names of things in the universe leaves you unfulfilled, then this book will provide insight to the deeper physical principles that underlie "astronomy as a science" rather than "astronomy as pretty pictures." The author is not afraid to use math, but unlike many other relatively advanced books, it does not use math as a substitute for communication with words.

General Physics

Cox, P. A. *The Elements: Their Origin, Abundance, and Distribution.* New York: Oxford University Press, 1989. I cannot add much to the book's title.

Peierls, Rudolf. *Surprises in Theoretical Physics.* Princeton: Princeton University Press, 1979. Many scientific discoveries are not planned. Here is an account of some of them.

Schwartz, Cindy. *A Tour of the Subatomic Zoo: A Guide to Particle Physics.* New York: American Institute of Physics, 1992. A heavily illustrated introduction to the abstract world of particle physics.

Swenson, Jr., Loyd S. *The Ethereal Aether.* Austin: University of Texas Press, 1972. Direct from the theory graveyard, this book documents the rise and fall of the concept of a luminiferous ether.

Weinberg, Steven. *Subatomic Particles.* Scientific American Library. New York: W. H. Freeman, 1983. A readable and somewhat anecdotal history of the discovery of atomic particles.

General Reference

Abbott, David. *The Biographical Dictionary of Scientists: Astronomers.* New York: Peter Bedrick Books, 1984. This one has everybody in it—from the famous to the obscure.

Jerrard, H. G., and D. B. McNeill. *Dictionary of Scientific Units.* London: Chapman and Hall, 1980. A useful reference for that moment when your memory of units does not measure up.

Mackay, Alan L. *Dictionary of Scientific Quotations.* New York: Adam Hilger, 1991. As complete a collection as you are likely to find. Includes quotes from politicians, novelists, and of course scientists. Entertaining and revealing.

Weast, Robert C. *CRC Handbook of Chemistry and Physics.* Cleveland: CRC Press, 1992. Accept no substitutes. Nearly four inches thick and weighing in at over five pounds, this reference book belongs on the shelf of all serious students of physics or chemistry.

General Science

Aveni, Anthony. *Empires of Time: Calendars, Clocks, and Cultures.* New York: Basic Books, 1989. The history of the measurement of time. Highly readable and, at times, entertaining.

Brancazio, Peter J. *Sport Science.* New York: Simon and Schuster, 1985. Required reading for the scientifically inclined sports enthusiast. The physics of everything from the discus throw to a basketball shot.

Flaste, Richard, ed., and the Editors and Reporters of Science Times [science section of the *New York Times*]. *New York Times Book of Scientific Literacy.* New York, Random House, 1990. A collection of science facts that the science editors of the *New York Times* have determined to be essential knowledge for the masses.

Gardner, Martin. *Fads and Fallacies in the Name of Science.* New York: Dover, 1957. An account of how people have claimed to use principles of science as a basis for fads, cults, and the paranormal. While many examples were especially relevant in the 1950s, the book still serves as a blueprint to help combat contemporary pseudoscience.

Getting Around the Sky

Allen, Richard Hinkley. *Star Names: Their Lore and Meaning.* New York: Dover, 1963. First published in 1899, this book contains more information about star names and constellations than you could have ever imagined. Its high density of information makes it better for the reference shelf than for the beach.

Astronomical Almanac, Superintendent of Documents, U.S. Government Printing Office, Washington, D.C., 20402. Published annually. If you are *really* serious about stargazing, then the *Astro-*

nomical Almanac is indispensable. It contains everything you never needed to know about cosmic happenings, from detailed eclipse timings to the rising and setting times of the Sun, Moon, and planets, to the differential coordinates of Jupiter's satellites. Indispensable to the professional astronomer and the committed amateur.

Burnham's Celestial Handbook: An Observer's Guide to the Universe Beyond the Solar System. 3 vols. New York: Dover, 1978. Alphabetical by constellation, these books contain over two thousand pages of readable information about everything that is worth your attention in the sky.

Norton, Arthur P. *Norton's Star Atlas and Reference Handbook.* Edited by Ian Ridpath. New York: Halsted Press, 1989. First published in 1910, this time-honored classic handbook of the heavens offers useful and informative chapters on telescopes and general astronomy. It also contains star charts with interesting objects clearly identified. Unlike Burnham's handbook, this one fits easily into your bag or briefcase.

Ottewell, Guy. *Astronomical Calendar.* Sky Publishing Company, P.O. Box 9111, Belmont, Massachusetts 02178-9111. Published annually. If you are serious about stargazing, then this one is for you. It gives a detailed account of cosmic happenings for every day of the year.

Seidelman, P. Kenneth, ed. *Explanatory Supplement to the Astronomical Almanac.* Mill Valley, Calif.: University Science Books, 1992. This is what you read when you get lost reading the *Astronomical Almanac.* It contains especially informative discussions of the reckoning of time, coordinate systems, and how the tables were computed for the *Astronomical Almanac.*

Voyager II: The Dynamic Sky Simulator for the Macintosh. Carina Software, 830 Williams Street, San Leandro, California 94577. If your local planetarium is too far away from you, then this is the next best thing. The software enables you to do things you have never even thought of (such as monitor the paths of the planets in the sky as they would appear if you lived on Pluto). With a color monitor, the program may be more interesting than actually going outside and looking up.

History of Science

Aristotle. *The Basic Works of Aristotle.* Edited by Richard McKeon. New York: Random House, 1941. Aristotle enjoyed less success in astronomy than in other pursuits of knowledge. This volume contains his most important and influential writings.

Einstein, Albert. *The Principle of Relativity: Collection of Original Papers on the Special and General Theory of Relativity.* New York: Dover, 1952. The original papers, including the one where $E = mc^2$ appears for the first time.

Galileo Galilei. *Dialogues Concerning Two New Sciences.* Translated by Henry Crew and Alfonso de Salvio. New York: Dover, 1954. A reprint of Galileo's original work from 1638. To convey his scientific discoveries to the lay reader, Galileo published his research on mechanics in the form of continuous dialogue between a teacher, Salviati, and two inquisitive students, Sagredo and Simplicio.

Newton, Sir Isaac. *Principia.* Vol. 1, *The Motion of Bodies.* Vol. 2, *The System of the World.* Berkeley: University of California Press, 1962. Reprinted from the original *Philosophiae Naturalis Principia Mathematica,* published in 1687. Newton's insight to the universe was staggeringly deep.

Shamos, Morris H. *Great Experiments in Physics: Firsthand Accounts from Galileo to Einstein.* New York: Dover, 1959. The original descriptions of the most significant experiments ever conducted in the history of physics.

Mathematics

Boyer, Carl B. *A History of Mathematics.* New York: Wiley, 1991. Concise, complete, and readable.

Kasner, Edward, and James R. Newman. *Mathematics and the Imagination* (originally published in 1940). Reprint. Redmond, Wash.: Tempus Books, 1989. The most enjoyable book I have ever read on mathematics. I first read it at age fourteen, and it is still on my shelf.

Smith, David Eugene, ed. *A Source Book in Mathematics*. New York: Dover, 1959. Original papers from the greatest thinkers in the history of mathematics.

Philosophy of Science

Feynman, Richard. *The Character of Physical Law*. Cambridge: MIT Press, 1965.

Heisenberg, Werner. *Physics and Philosophy: The Revolution in Modern Science*. New York: Harper and Row, 1958.

Kuhn, Thomas. *The Structure of Scientific Revolution*. 2d ed. Chicago: University of Chicago Press, 1970. These are three timeless classics. They are short and mostly accessible to a general audience.

Assorted

Edelman, Bernard, *Dear America: Letters Home from Vietnam*. New York: W. W. Norton, 1985. Revealing and often upsetting first-hand descriptions of combat during the Vietnam War. (Used for parts of chapter 6.)

Major, Frederick, M. C. Myatt, and Gerard Ridefort. *Modern Rifles and Submachine Guns*. London: Salamander Books, 1992. Detailed yet brief discussions of nearly every important rifle and machine gun used in warfare. (Used for parts of chapter 6.)

Shakespeare, William. *Riverside Shakespeare*. Edited by G. Blakemore Evans. Boston: Houghton Mifflin, 1974. A complete collection of Shakespeare's plays and poems, with supplemental material on the life and times of Shakespeare himself. (Used for parts of chapters 4 and 13.)

Index

Note: Page numbers in italics refer to figures.

Text: Palatino

Compositor: Maple-Vail

Printer: Maple-Vail

Binder: Maple-Vail